Silke Alt

New aminocoumarin antibiotics from genetic engineering

AF062241

Silke Alt

New aminocoumarin antibiotics from genetic engineering

Their development and biological activities

Südwestdeutscher Verlag für
Hochschulschriften

Imprint
Any brand names and product names mentioned in this book are subject to trademark, brand or patent protection and are trademarks or registered trademarks of their respective holders. The use of brand names, product names, common names, trade names, product descriptions etc. even without a particular marking in this work is in no way to be construed to mean that such names may be regarded as unrestricted in respect of trademark and brand protection legislation and could thus be used by anyone.

Publisher:
Südwestdeutscher Verlag für Hochschulschriften
is a trademark of
Dodo Books Indian Ocean Ltd., member of the OmniScriptum S.R.L Publishing group
str. A.Russo 15, of. 61, Chisinau-2068, Republic of Moldova Europe
Printed at: see last page
ISBN: 978-3-8381-2661-6

Zugl. / Approved by: Tübingen, Universität, Diss., 2010

Copyright © Silke Alt
Copyright © 2011 Dodo Books Indian Ocean Ltd., member of the OmniScriptum S.R.L Publishing group

New aminocoumarin antibiotics from genetic engineering

Their development and biological activities

Neue Aminocoumarin-Antibiotika: Gentechnische Herstellung und biologische Testung

CONTENTS

ABBREVIATIONS		5
SUMMARY		7
ZUSAMMENFASSUNG		9
I.	INTRODUCTION	12
I.1.	Antibiotic resistance and the quest for new antibiotics	12
I.2.	Aminocoumarin antibiotics	14
I.2.1.	Chemical structure	14
I.2.2.	Mechanism of action	15
I.2.3.	Clinical application	17
I.2.4.	Biosynthesis and biosynthetic gene clusters	18
I.2.5.	Bacterial resistance mechanisms against aminocoumarins	20
I.2.6.	Transport of catechol siderophores	22
II.	MATERIALS AND METHODS	24
II.1.	Microbiology methods	24
II.1.1.	Microorganisms	24
II.1.2	Culture media	25
II.1.3.	Growth and preservation of microorganisms	27
II.1.4	Antibiotic solutions	28
II.2.	Molecular biology methods	29
II.2.1.	Vectors and constructs	29
II.2.2.	DNA isolation	30
II.2.3.	DNA quantification and manipulation with enzymes	32
II.2.4.	PCR amplification of DNA	33
II.2.5.	Agarose gel electrophoresis of DNA	34
II.2.6.	Introduction of DNA in *E. coli* and *Streptomyces*	35
II.2.7.	Construction and heterologous expression of plasmid pSA11	37
II.2.8.	Construction and heterologous expression of cosmid clo-SA2	37

II.2.9.	Construction and heterologous expression of cosmid clo-SA4	38
II.2.10.	Generation of *E. coli* mutants	39

II.3. Biochemistry methods 39

II.3.1.	Assay compounds, enzymes, DNAs and chemicals	39
II.3.2.	General methods for protein expression and purification	40
II.3.3.	Cloning, protein expression, and purification of *S. aureus* topoisomerase IV subunits ParC and ParE	41
II.3.4.	Cloning, protein expression, and purification of the AMP ligase DhbE	42
II.3.5.	Denaturating polyacrylamide gel electrophoresis (SDS PAGE) and Coomassie staining	42
II.3.6.	Amide synthetase assay	44
II.3.7.	Topoisomerase IV decatenation assay	44
II.3.8.	DNA gyrase supercoiling assay	45
II.3.9.	Agar diffusion test	46

II.4. Analytical chemistry techniques 47

II.4.1.	Production and purification of novclobiocin 401	47
II.4.2.	HPLC analysis	47
II.4.3.	LC-MS analysis	48
II.4.4.	NMR analysis	48

III. RESULTS 49

III.1. Generation and activity test of novclobiocin 401, a clorobiocin derivative containing the catechol moiety 3,4-dihydroxybenzoic acid 49

III.1.1.	Investigation of the substrate tolerance of different aminocoumarin acyl ligases for acyl substrates with catechol moieties	49
III.1.2.	Inactivation of *cloQ* in the biosynthetic gene cluster of clorobiocin, and heterologous expression of the modified cluster	51
III.1.3.	Mutasynthetic experiments with 3,4-DHBA and caffeic acid	52
III.1.4.	Creating an artificial pathway to 3,4-DHBA	52
III.1.5.	Production of novclobiocin 401 by *Streptomyces coelicolor*(clo-SA2) harbouring plasmid pSA11	54
III.1.6.	Structure elucidation of novclobiocin 401	55

III.1.7.	Inhibitory activities against *E. coli* and *S. aureus* DNA gyrase and topoisomerase IV	58
III.1.8.	Construction of *E. coli* mutants for investigation of antibiotic import by catechol siderophore transporters	60
III.1.9.	Determination of the antibacterial activity of novclobiocin 401 in agar diffusion tests	62
III.1.10.	Growth promotion with enterobactin	65
III.1.11.	Determination of the minimum inhibitory concentration (MIC) of novclobiocin 401	66
III.2.	**Generation of a clorobiocin derivative containing the catechol moiety 2,3-dihydroxybenzoic acid**	**67**
III.2.1.	Activation of 2,3-dihydroxybenzoic acid by the AMP ligase DhbE from *Bacillus subtilis*	67
III.2.2.	Detection of the clorobiocin derivative with 2,3-dihydroxybenzoic acid	70
III.3.	**Inhibition of DNA gyrase and topoisomerase IV of *S. aureus* and *E. coli* by aminocoumarin antibiotics**	**72**
III.3.1.	Expression of the subunits of *S. aureus* topoisomerase IV as his-tagged proteins	72
III.3.2.	Removal of potassium glutamate from the assay for DNA gyrase activity	72
III.3.3.	Effect of potassium glutamate on the activity of DNA gyrase and topoisomerase IV of *E. coli* and *S. aureus*	73
III.3.4.	Potassium glutamate modulates the sensitivity of *E. coli* DNA gyrase to aminocoumarin antibiotics	76
III.3.5.	Inhibition of DNA gyrase and topoisomerase IV from *E. coli* and *S. aureus* by different aminocoumarin antibiotics	79
III.4.	**Inactivation of *cloHIJK* in the biosynthetic gene cluster of clorobiocin and heterologous expression of the modified cluster**	**81**
IV.	**DISCUSSION**	**83**
IV.1.	Generation and activity test of novclobiocin 401, a clorobiocin derivative containing the catechol moiety 3,4-dihydroxybenzoic acid	83
IV.2.	Inhibition of DNA gyrase and topoisomerase IV of *S. aureus* and *E. coli* by aminocoumarin antibiotics	86
V.	**REFERENCES**	**89**

ABBREVIATIONS

°C	degree Celsius
µ	micro
aac(3)IV	apramycin acetyltransferase (apramycin resistance gene)
aadA	streptomycin and spectinomycin resistance gene
Amp	ampicillin
Apr	apramycin
ATP	adenosine triphosphate
bp	base pair
B. subtilis	*Bacillus subtilis*
CFU	colony forming unit
Cm	chloramphenicol
Da	dalton
DHBA	dihydroxybenzoic acid
DMSO	dimethyl sulfoxide
DNA	deoxyribonucleic acid
dNTP	deoxyribonucleoside 5`-triphosphate
DTT	1,4-dithiothreitol
E. coli	*Escherichia coli*
EDTA	ethylenediamine tetra-acetic acid
g	gram
GyrA	DNA gyrase subunit A
GyrB	DNA gyrase subunit B
h	hour
HCl	hydrochloric acid
HCOOH	formic acid
His$_6$	hexahistidine
HPLC	high performance liquid chromatography
Hz	hertz
IPTG	isopropyl-β-thiogalactoside
k	kilo
Km	kanamycin
K-Glu	potassium glutamate
kb	kilobase pair
kDa	kilodalton
kDNA	kinetoplast DNA
l	litre
M	molar
m	milli
min	minute
MS	mass spectrometry
MW	molecular weight
n	nano
NaCl	sodium chloride
Na-Glu	sodium glutamate
NaOH	sodium hydroxide
neo	aminoglycoside phosphotransferase (kanamycin resistance gene)
nm	nanometre

ABBREVIATIONS

NMR	nuclear magnetic resonance
OD$_{600}$	optical density at 600 nm
ORF	open reading frame
oriT	origin of transfer from RK2
p	pico
ParC	topoisomerase IV subunit C
ParE	topoisomerase IV subunit E
PCR	polymerase chain reaction
PEG	polyethylene glycol
r (superscript)	resistant
ref.	reference
Ring A	3-dimethylallyl-4-hydroxybenzoic acid
RNase	ribonuclease
RP	reverse phase
rpm	revolutions per minute
RT	room temperature
s (superscript)	sensitive
s	second
S. coelicolor	*Streptomyces coelicolor*
S. aureus	*Staphylococcus aureus*
SAM	S-Adenosyl methionine
SDS	sodium dodecyl sulphate
SDS-PAGE	sodium dodecyl sulphate-polyacrylamide gel electrophoresis
Str	streptomycin
TEMED	N, N, N`,N`-tetramethylethylenediamine
TES	N-Tris-(hydroxymethyl)-methyl-2-aminoethanesulfonic acid
Tet	tetracycline
Tsr	thiostrepton
Tris	2-amino-2-(hydroxymethyl)- 1,3-propanediol
Topo	topoisomerase
U	unit
UV	ultraviolet
WT	wild-type

SUMMARY

The increasing resistance of pathogenic bacteria to existing antibiotics has become one of the most serious threats to public health, and the discovery and development of new antibiotics represents an enormous challenge both for industry and for academic research institutions.

Aminocoumarin antibiotics like clorobiocin, novobiocin and coumermycin A_1 are produced by different *Streptomyces* strains. Their biosynthetic gene clusters have been cloned and sequenced, and the function of nearly all genes therein has been elucidated. With this knowledge, new derivatives of these antibiotics were generated by genetic engineering, mutasynthesis and combinatorial biosynthesis over the past years. In contrast to fluoroquinolones, aminocoumarins are potent inhibitors of the GyrB subunit of the bacterial DNA gyrase by competing with the binding of ATP. They are active against Gram-positive pathogens including methicillin-resistant *Staphylococcus aureus* strains and have also potential applications in oncology. It has been shown that they enhance the cytotoxic activities of the anti-tumor drugs etoposide and teniposide and interact with the eucaryotic heat shock protein 90 (Hsp90). Only novobiocin had been licensed for clinical use in human infections in the United States (Albamycin®). Because of the low solubility in water, toxicity in eukaryotes and poor penetration in Gram-negative bacteria the therapeutic use of aminocoumarin antibiotics remain restricted. Combinatorial biosynthesis may offer a way to develop novel aminocoumarins with improved properties.

Previous date indicated that the lack of antibiotic activity of aminocoumarins against Gram-negative bacteria was in part due to their poor permeation across the outer membrane. In this work, we aimed to generate a siderophore-like derivative of clorobiocin, which mimicks the structure of siderophores. This would facilitate the active transport of the antibiotic into the cell by its own siderophore transporters. In the first part of this PhD work, the prenylated 4-hydroxybenzoyl moiety (Ring A) of clorobiocin was replaced with a 3,4-dihydroxybenzoyl moiety using combinatorial biosynthesis techniques. An artificial operon was synthesized, consisting of the genes for chorismate pyruvate lyase of *E. coli* and for 4-hydroxybenzoate 3-hydroxylase of *Corynebacterium cyclohexanicum*. This operon, directing the biosynthesis of 3,4-dihdroxybenzoate, was expressed in the heterologous host *Streptomyces coelicolor* M512, together with a modified clorobiocin gene cluster that lacks an essential gene for the biosynthesis of the genuine Ring A. The resulting strain now produced a new clorobiocin derivative containing a 3,4-dihdroxybenzoyl moiety. Its structure was confirmed by LC-MS, HR-MS and NMR analysis, and it was found to be a potent inhibitor of the DNA gyrase from *E. coli* and *Staphylococcus aureus*. These experiments confirmed that the structure of Ring A is not essential for the interaction with DNA gyrase. Bioassays against different *E. coli* mutants suggested that this compound (novclobiocin 401) was actively imported by catechol siderophore transporters in the cell envelope. This study provides a new example that the structure of a natural product can be rationally modified by genetic methods.

Experiments to generate other siderophore-like derivatives of clorobiocin were performed. Substitution of Ring A with 2,3-dihydroxybenzoic acid (2,3-DHBA), which is the catechol moiety present in the siderophore enterobactin of *Escherichia coli*, was achieved at the end of this work. The limiting biosynthetic steps were identified by *in vitro* assays: 2,3-DHBA must be first activated by adenylation –attained with the AMP ligase DhbE from *Bacillus subtilis*– before being accepted by the available aminocoumarin acyl ligases as substrate.

In another project, the inhibitory activity of the naturally occurring aminocoumarin antibiotics novobiocin, clorobiocin, coumermycin A_1, simocyclinone D8 and of several new derivatives (novclobiocins) against DNA gyrase and topoisomerase IV from *Escherichia coli* and *Staphylococcus aureus* was investigated as well as the effect of potassium and sodium glutamate on the activity of these enzymes. For this purpose, the inhibitory concentrations of the aminocoumarins were determined in DNA gyrase supercoiling assays and topoisomerase IV decatenation assays. Both subunits of *S. aureus* topoisomerase IV were purified by expressing the genes encoding the subunits ParC and ParE separately as His-Tag proteins in *Escherichia coli*. DNA gyrase is *in vitro* the primary target of all investigated aminocoumarins. With the exception of simocyclinone D8, all other aminocoumarins inhibited *S. aureus* DNA gyrase on average 6-fold more effectively than *E. coli* DNA gyrase. Potassium glutamate was found to be essential for the activity of *S. aureus* DNA gyrase and increased the sensitivity of *E. coli* DNA gyrase to aminocoumarins at least 10-fold. Furthermore, the IC_{50} values were three orders of magnitude lower than those reported for fluoroquinolones. This study provides insights about the important substituents for the inhibitory activity of aminocoumarins against the target enzymes, and thereby may facilitate the rational design of improved antibiotics.

ZUSAMMENFASSUNG

Die steigende Resistenz pathogener Bakterien gegenüber bekannten Antibiotika wird immer mehr zu einer ernstzunehmende Bedrohung der Gesundheit und die Entdeckung und Erforschung neuer wirksamer Antibiotika zu einer schwierigen Herausforderung.
Aminocoumarine, wie z.b. Novobiocin, Clorobiocin oder Coumermycin A_1, bilden eine interessante Gruppe von Antibiotika, die von verschiedenen Stämmen der Gattung *Streptomyces* gebildet werden. Ihre Biosynthesegencluster wurden in den vergangenen Jahren kloniert, sequenziert und die Funktion nahezu aller Gene aufgeklärt. Mit diesem Wissen konnten neue Derivate dieser Antibiotika durch genetische Manipulation, Mutasynthese und Kombinatorische Biosynthese hergestellt werden.
Im Gegensatz zu Chinolon-Antibiotika, die an der A-Untereinheit der bakteriellen Gyrase angreifen, ist das zelluläre Traget der Aminocoumarine die B-Untereinheit. Das therapeutische Potential der Aminocoumarine liegt in ihrer hohen Affinität zur bakteriellen Gyrase mit Inhibierungskonzentrationen im 10 nM Bereich; d.h. die Hemmkonzentrationen sind erheblich geringer als die der Chinolone.
Neben ihrer Wirkung als Antiinfektiva gegen Gram-positive Pathogene (methicillin-resistente *Staphylococcus aureus* Stämme eingeschlossen) finden sie außerdem Anwendung in der Onkologie. Aminocoumarine potenzieren zum einen die zytotoxische Wirkung der Topoisomerase-Inhibitoren Etoposid und Teniposid und reduzieren zum anderen durch direkte Interaktion mit dem Heat shock Protein 90 (Hsp90) die Menge an onkogenen Protein Kinasen (z.B. Raf-1) und somit die Anzahl an Tumorzellen.
Obwohl Clorobiocin der potentere Wirkstoff gegen die bakterielle Gyrase ist, ist einzig Novobiocin in den USA als humantherapeutisches Antiinfektivum unter dem Handelsnamen Albamycin® (Pharmacia & Upjohn) zugelassen und wird zur Behandlung multiresistenter Gram-positiver Pathogene wie *Staphylococcus aureus* und *Staphylococcus epidermidis* eingesetzt. Wegen der schlechten Löslichkeit in Wasser, der Toxizität gegenüber eukaryotischen Zellen und der geringen Aktivität gegen Gram-negative Bakterien, wird Novobiocin allerdings nur als Reserveantibiotikum verwendet. Kombinatorische Biosynthese bietet daher eine viel versprechende Möglichkeit neue Aminocoumarin-Antibiotika mit verbesserten Eigenschaften zu entwickeln.

ZUSAMMENFASSUNG

Zu Beginn dieser Arbeit wurde angenommen, dass die schlechte Wirkung von Aminocoumarinen auf die unzulängliche Penetration durch die äußere Membran Gram-negativer Bakterien zurückzuführen ist. Wir stellten daraufhin mittels Kombinatorischer Biosynthese ein Clorobiocin Derivat mit einer siderophor-ähnlichen Struktureinheit her, mit deren Hilfe der aktive Transport durch die äußere Membran mittels zelleigenen Siderophortransporter genutzt werden sollte. Hierfür wurde die prenylierte 4-Hydroxybenzoesäure (Ring A) von Clorobiocin durch 3,4-Dihydroxybenzoesäure (3,4-DHBS) ersetzt. Zuvor wurde ein Operon aus zwei synthetischen Genen erstellt, die für eine Pyruvatlyase und eine 4-Hydroxybenzoat-3-hydroxylase kodieren und zusammen für die Biosynthese der 3,4-DHBS verantwortlich sind. Die Gene stammen aus dem Gram-negativen Bakterium *E. coli* und dem Gram-positiven Bakterium *Corynebacterium cyclohexanicum*. Die Codons beider Gensequenzen wurden für eine Expression im Wirtsstamm *Streptomyces coelicolor* M512 optimiert und zusammen mit einem *cloQ*-defekten Clorobiocin Biosynthesegencluster (verantwortlich für das Ausbleiben der Ring A-Biosynthese) transformiert. Der resultierende Stamm produzierte ein neues Clorobiocin Derivat (Novclobiocin 401) mit der siderophor-ähnlichen Struktureinheit 3,4-DHBS. Die chemische Struktur des neuen Clorobiocin Derivats wurde mittels LC-MS, HR-MS und NMR Analysen bestätigt.

Novclobiocin 401 erwies sich als potenter Hemmstoff der bakteriellen Gyrase von *Escherichia coli* und *Staphylococcus aureus*, was die Vermutung bestätigte, dass Ring A nicht an der direkten Interaktion mit der bakteriellen Gyrase beteiligt ist. Bioassays gegen verschiedene *Escherichia coli* Mutanten zeigten, dass diese Substanz aktiv durch Catecholsiderophortransporter in die Zelle transportiert wurde. Anhand dieser Arbeit konnte gezeigt werden, dass neue potente Substanzen bewusst durch „synthetische Biologie" hergestellt werden können.

Im Laufe dieser Arbeit wurde außerdem erstrebt, andere siderophor-ähnliche Clorobiocin Derivate herzustellen. Eine Substitution von Ring A mit 2,3-Dihydroxybenzoesäure (2,3-DHBS), der Catecholeinheit der Siderophore Enterobactin von *Escherichia coli*, wurde schlussendlich erhalten, nachdem der limitierende Schritt durch *in vitro* Assays identifiziert wurde: 2,3-DHBS musste erst durch Adenylierung aktiviert werden, was durch Inkubation mit der AMP Ligase DhbE aus *Bacillus subtilis* erreicht wurde. Erst die aktivierte 2,3-DHBS wurde danach von allen uns vorliegenden Aminocoumarin-Acyl Ligase als Substrat akzeptiert.

ZUSAMMENFASSUNG

Im zweiten Projekt wurde die inhibitorische Aktivität der natürlichen Aminocoumarine Clorobiocin, Novobiocin, Coumermycin A_1, Simocyclinon D8 und mehrere Aminocoumarin Derivate (Novclobiocine) gegen die bakterielle Gyrase und Topoisomerase IV von *Escherichia coli* und *Staphylococcus aureus* untersucht. Außerdem wurde der Effekt von Kaliumglutamat (K-Glu) auf die Aktivität der Topoisomerasen bestimmt. Die Inhibierungskonzentartionen der Aminocoumarine wurden in Gyrase-Supercoiling Assays bzw. Topoisomerase IV-Dekatenierungs Assays bestimmt. Beide Untereinheiten der *S. aureus* Topoisomerase IV, ParC und ParE, wurden exprimiert und mittels Nickel-Affinitätschromatographie gereinigt. Die bakterielle Gyrase erwies sich in den *in vitro* Assays als bevorzugte Zielstruktur aller getesteten Aminocoumarine. Mit Ausnahme von Simocyclinon D8 wurde die *S. aureus* gyrase im Durchschnitt sechsmal besser als die *E. coli* Gyrase gehemmt. Kaliumglutamat erwies sich als unabkömmlich für die Aktivität der *S. aureus* Gyrase und steigerte die Sensitivität der *E. coli* Gyrase mindestens um das 10-fache. Die Ergebnisse dieses Projektes zeigten die Wichtigkeit verschiedener Substituenten der Aminocoumarin Derivate für die Interaktion mit ihren Zielstrukturen, was für das Design zukünftiger potenter Aminocoumarine von Wichtigkeit sein kann.

I. INTRODUCTION

I.1. Antibiotic resistance and the quest for new antibiotics

The dramatic increase in the number of antibiotic-resistant pathogenic Gram-positive and Gram-negative bacteria in the past decade have focused attention on the need for new anti-infective drugs[119]. Natural products have traditionally played a dominant role in the discovery of new drugs, although structural modifications of these natural products are often necessary to improve the efficacy, stability and pharmacokinetics[124].

About two-thirds of the known antibiotics, and many other compounds with different biological activities, are produced by a group of Gram-positive bacteria called actinomycetes[9], characterised by the high content of guanine and cytosine of their genomes[125] (in contrast with the other main group of Gram-positive bacteria, the low G+C organisms, which includes genera such as *Bacillus* or *Staphylococcus*). Among actinomycetes, the most prolific producers of antibiotics are the species of the genus *Streptomyces*. *Streptomyces* are ubiquitous soil bacteria, where they play a central role in carbon recycling. They show one of the most complex life cycles among bacteria: a spore germinates leading to the development of a vegetative mycelium that differentiates in aerial hyphae and later in chains of spores, facilitating the dispersion of the specie as well as its survival under adverse conditions. The production of antibiotics is usually coordinated with the morphological differentiation[59].

The history of antibiotics obtained from *Streptomyces* begins with the discovery of streptothricin in 1942, although it was the discovery of streptomycin two years later what triggered systematic screening of antibiotics produced by members of this genus[122]. *Streptomyces coelicolor* A3(2) is the most studied representative of this genus. This strain is able to produce four known chromosomally encoded antibiotics: the NRPS/PKS-derived prodiginines, the type II polyketide actinorhodin, the NRPS-derived calcium-dependent antibiotic (CDA) and the type I polyketide CPK[11].

Antibiotic production in *Streptomyces* is generally growth phase-dependent. In liquid culture it begins as the culture enters stationary phase. Most antibiotics are the products of complex biosynthetic pathways, with all the enzymes and transport proteins encoded within a gene cluster that usually contains also genes for pathway-specific transcriptional regulatory proteins. The onset of antibiotic biosynthesis is determined and influenced by a variety of physiological and environmental factors. There is evidence that the full capacity for secondary metabolite production by soil microorganisms is not expressed under the typical conditions used for antibiotic screening in the laboratory. Furthermore, only a small portion of microorganisms are culturable by current methods. Therefore, it is accepted that a large number of biosynthetic pathways still await to be discovered in actinomycetes, what encourages the use of genetic approaches for antibiotic discovery, like heterologous expression of biosynthetic gene clusters in culturable and genetic manipulable strains (e.g. *Streptomyces coelicolor* M512[38] and combinatorial biosynthesis or genetic engineering as methods to increase the natural diversity[120].

In 1985, Hopwood[55] reported 1985 for the first time the production of "hybrid" antibiotics by genetic engineering of *Streptomyces* strains, i.e. through the transfer of biosynthetic genes between strains producing different members of the same class of antibiotics in order to combine structural features of both compounds. Since then, the number of biosynthetic genes and gene clusters available for such experiments, and the genetic techniques available for recombination and expression, have expanded greatly and many new bioactive compounds have been generated by genetic engineering of microorganisms[53, 54]. Advances in the methods of chemical DNA synthesis now allow to readily adapt the sequence of a gene to different expression hosts, what greatly expands the possibilities for the generation of new bioactive compounds by the combination of genes from very different organisms. This "combinatorial biosynthesis", or the shuffling of biosynthetic genes from different pathways and even organisms via genetic engineering to create novel chemical structures, has proved in the last few years to represent a promising alternative approach to create new compounds and overcome bacterial resistance to existing drugs[21, 76, 119].

Over the last years, the biosynthetic gene clusters of five different aminocoumarin antibiotics have been cloned and sequenced, and this group of antibiotics has become a successful example for the generation of new derivatives in high structural diversity by combinatorial biosynthesis techniques, including metabolic engineering, mutasynthesis, and chemoenzymatic synthesis in natural and heterologous producer strains[65, 35, 48].

I.2. Aminocoumarin antibiotics

I.2.1. Chemical structure

The aminocoumarin antibiotics are produced by different *Streptomyces* strains: e.g. novobiocin by *S. spheroides*[111], clorobiocin by *S. roseochromogenes* var. *oscitans* DC 12.976[72], and coumermycin A_1 by *S. rishiriensis*[13]. So far, two further aminocoumarin antibiotics, simocyclinone D8[101] and rubradirin[109] have been discovered **(Figure I.1)**. The characteristic structural moiety that gives the name to the aminocoumarin antibiotics is a 3-amino-4,7-dihydroxycoumarin moiety (Ring B), which is linked via an amide bond to an acyl moiety (Ring A) and via a glycosidic bond to the deoxysugar noviose (Ring C)[49]. Clorobiocin differs from novobiocin at two positions: novobiocin has a carbamoyl group attached to noviose, while clorobiocin contains a 5-methylpyrrole-2-carboxylic acid molecule, and clorobiocin carries a chlorine atom at Ring B whereas novobiocin has a methyl group at the corresponding position **(Figure I.1)**.

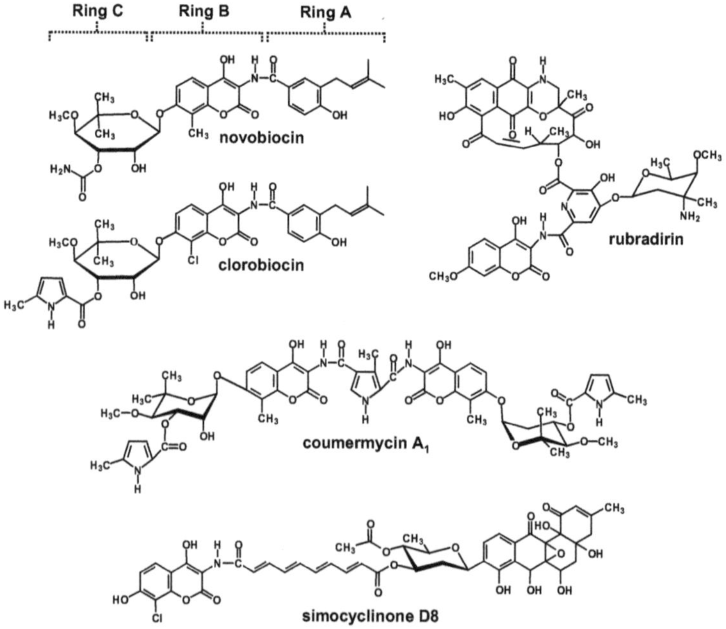

Figure I.1:
Chemical structure of the aminocoumarin antibiotics.

Coumermycin A_1 contains two noviosyl aminocoumarin moieties and has a different acyl component, 3-methyl-pyrrole-2,4-dicarboxylic acid. Simocyclinone D8 and rubradirin have only Ring B in common with the "classical" aminocoumarins[49].

I.2.2. Mechanism of action

The mode of action of clorobiocin, novobiocin and simococlinone D8 has been examined recently by Maxwell and Lawson[74]. These aminocoumarins are powerful inhibitors of DNA gyrase, binding to this target with higher affinity than modern fluoroquinolones.
Gyrase and topoisomerase IV belong to the procaryotic type II topoisomerases[22, 104], that are different from eukaryotic topoisomerases and therefore they are a promising anti-infective drug target. Topoisomerases can be divided into two main classes: type I enzymes, which cleave a single strand of DNA during the course of their reaction, and type II enzymes, which cleave both strands. DNA gyrase is unique in its ability to introduce negative supercoils into DNA and is involved in the maintenance of a critical superhelical density of DNA that is essential for DNA replication and transcription.

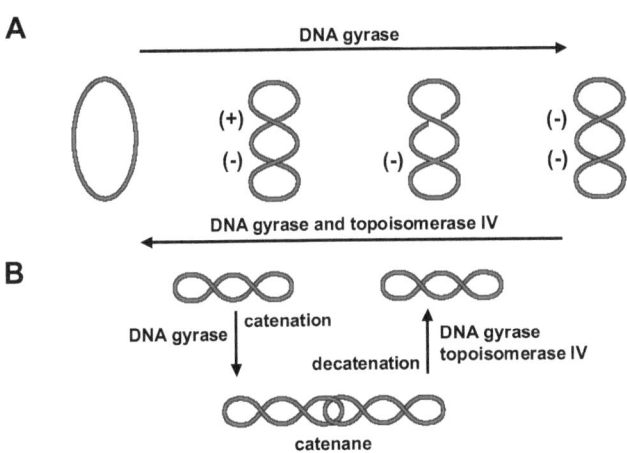

Figure I.2[107]:
Reactions of type II topoisomerases. **(A)** DNA gyrase introduces negative supercoils into closed circular DNA by the concerted breaking and rejoining of double strands. Both DNA gyrase and topoisomerase IV can remove supercoils. **(B)** Supercoiled DNA is catenated by DNA gyrase and decatenated by both DNA gyrase and topoisomerase IV.

In contrast, the primary function of topoisomerase IV is the decatenation of multiply linked daughter chromosomes during the terminal stages of DNA replication **(Figure I.2)**. DNA gyrase as well as topoisomerase IV consist of two GyrA/GyrB and two ParC/ParE subunits, respectively. The reactions catalysed by both enzymes are energetically driven by hydrolysis of ATP, catalysed by GyrB and ParE subunits[22, 104].

Aminocoumarins bind to the GyrB subunit of DNA gyrase or the ParE subunit of topoisomerase IV, competing with the binding of ATP[10, 74]. Detailed crystallographic studies have been published on the interaction of clorobiocin and novobiocin with DNA gyrase and topoisomerase IV, which have shown that Ring B and Ring C are essential for the interactions with gyrase, while Ring A is much less involved in the binding of the antibiotics to the target. Therefore, it appears possible to vary the structure of Ring A without severely affecting the DNA gyrase inhibitory activity. Complexes between clorobiocin or novobiocin and the protein involve hydrophobic interactions and a network of hydrogen bonds. Key hydrogen bonds include those between Arg136 and Ring B, Asp73 and Thr65 and the acyl group of Ring C, and Asn46 and the hydroxyl group of Ring C **(Figure I.3)**[74]. The drugs do not occupy the same binding pocket as ATP, but the binding site of Ring C overlaps with the binding site for the adenine ring of ATP[86].

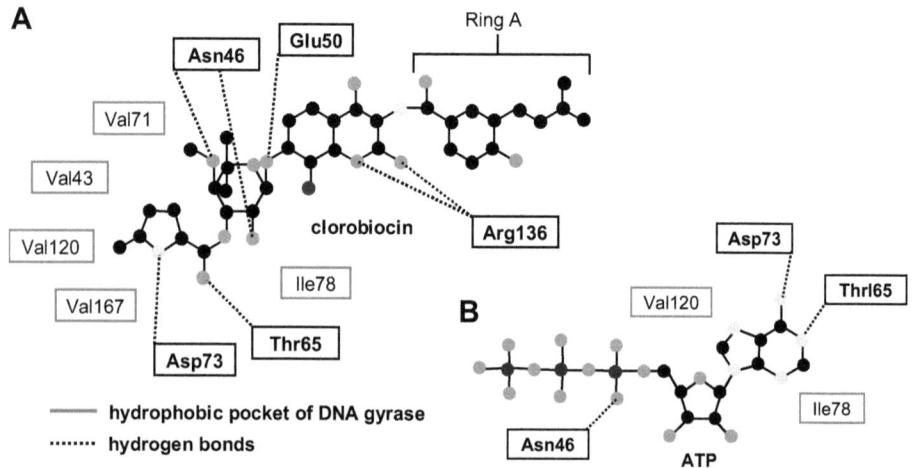

Figure I.3:

(A) Interaction between *E. coli* gyrase subunit B and clorobiocin; adapted from ref. 74; **(B)** Interaction between *E. coli* gyrase subunit B and ATP; adapted from ref. 86.

The interest in aminocoumarins has been stimulated by recent biochemical and X-ray crystallographic evidence showing that the aminocoumarin antibiotic simocyclinone D8 inhibits DNA gyrase by a completely new mode of action, interacting with two separate pockets of the enzyme and thereby preventing its binding to DNA[30, 36, 87].

I.2.3. Clinical application

The therapeutic potential of the aminocoumarins lies especially in their very high affinity to gyrase. Their equilibrium dissociation constants (K_D) for gyrase is in the 10 nM range, i.e. lower than that of fluoroquinolones[74].

Even though clorobiocin is a more potent inhibitor of DNA gyrase, only novobiocin (Albamycin®, Pharmacia & Upjohn) has been licensed in the United States for the treatment of human infections with multi resistant bacteria such as *Staphylococcus aureus* and *Staphylococcus epidermidis*[6]. Due to its poor solubility in water, which prevented the development of parenteral formulations, the toxicity in eukaryotes and its low activity against Gram-negative bacteria (resulting from poor permability), clinical use of this antibiotic remains restricted.

Novobiocin and its derivatives have also been investigated as potential anticancer drugs. Novobiocin acts synergistically with etoposide and teniposide and could be used in combination therapies to overcome drug resistance. The increase in etoposide cytotoxicity is due to the inhibition by novobiocin of etoposide efflux[68, 95-97]. Furthermore, novobiocin, clorobiocin, and coumermycin A_1 were shown to interact with the eukaryotic heat shock protein 90 (Hsp90), which plays a key role in the stability and function of multiple cell-signaling components, e.g. several oncogenic tyrosine and serine-threonine kinases (e.g. Raf-1), being expressed at two to ten fold higher levels in tumour cells than in their normal counterparts. Hsp90 is therefore considered to be a novel molecular target for anticancer therapeutics, and aminocoumarins markedly reduced cellular levels of oncogenic kinases *in vitro* and *in vivo* (mice) by interacting with Hsp90[73].

I.2.4. Biosynthesis and biosynthetic gene clusters

The biosynthesis of novobiocin was first studied in the 1960s by feeding experiments by Birch and co-workers[14] and Bunton and co-workers[20]. They proved that the deoxysugar is derived from glucose, and that tyrosine is the precursor of both the aminocoumarin moiety (Ring B) and the acyl component (Ring A). In 2000, cloning and sequencing of the novobiocin biosynthetic gene cluster provided the basis for genetic investigations of aminocoumarin antibiotics biosynthesis[111]. Subsequently, the biosynthetic gene clusters for coumermycin A_1 and clorobiocin were also identified[92, 121]. Comparison of the three gene clusters revealed a strikingly stringent correspondence between the structure of the antibiotics and the organization of the biosynthetic genes (**Figure I.4**). The order of the genes coding for each structural moiety are perfectly identical for the three clusters. The novobiocin, clorobiocin and coumermycin A_1 cluster span 23.4, 35.6 and 38.2 kb and comprise 20, 29 and 31 putative genes, respectively[66].

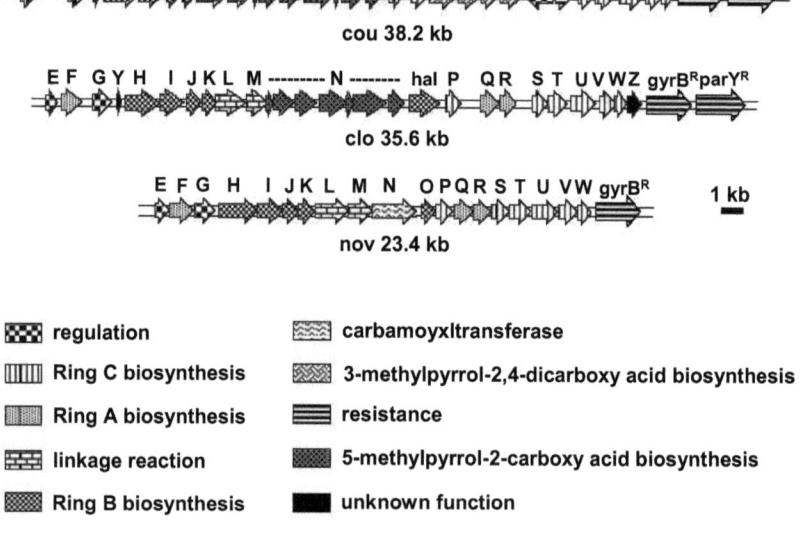

Figure I.4:
The biosynthetic gene clusters of coumermycin A_1 (cou), clorobiocin (clo) and novobiocin (nov).

INTRODUCTION

The 3-amino-4,7-dihydroxy coumarin moiety (Ring B) is present in all three aminocoumarins, and correspondingly all three clusters contain a group of four genes for its biosynthesis, i.e. *novHIJK*, *cloHIJK* and *couHIJK*[23].

The 3-prenylated-4-hydroxybenzoyl moiety (Ring A) of clorobiocin (and of novobiocin) is formed from 4-hydroxyphenylpyruvate under catalysis of the ABBA prenyltransferases CloQ[93] (NovQ) and two subsequent oxidative decarboxylation steps catalysed by the non-heme iron(II)-and α-ketoacid-dependent oxygenase CloR[91] (NovR). The prephenate dehydrogenases CloF (NovF) provides 4-hydroxyphenylpyruvate[34].

The aminocoumarin moiety is linked to the respective acyl moiety by the amide synthetase NovL[110], CloL[43] or CouL[103].

The aminocoumarin moieties of novobiocin and coumermycin A_1 are methylated by the methyl transferases NovO and CouO respectively[67]. Clorobiocin contains a chlorine atom and correspondingly the gene cluster of clorobiocin contains a gene *clo-hal* encoding a halogenase[32].

All three aminocoumarin antibiotics contain the same deoxysugar skeleton[92, 111, 121], i.e. 5-C-methyl-L-rhamnose, and all three gene clusters contain a group of five genes, *novSTUVW* and its orthologous in the other clusters. The dTDP-activated deoxysugar is subsequently transferred to the 7-hydroxy group of the aminocoumarin moiety, catalysed by the glycosyl transferases NovM[39], CloM and CouM.

After glycoside formation, the 4-hydroxy group of the deoxysugar is methylated by the SAM-dependent methyltransferases NovP[40], CloP or CouP.

The last step in the biosynthesis appears to be the acylation of the 3-hydroxy group of the deoxysugar, catalysed in novobiocin biosynthesis by the carbamoyl transferase NovN[40]. In clorobiocin and coumermycin A_1, the corresponding acyl moiety is a pyrrol-2-carboxylic acid formed by the genes *cloN1-7*[3] and *couN1-7* (**Figure I.5**). The clorobiocin and coumermycin A_1 gene clusters contain a small open reading frame, *cloY* and *couY*, which shows sequence similarity to the gene *mbtH* from the biosynthetic gene cluster of the siderophore mycobactin from *Mycobacterium tuberculosis*. It is supposed that these genes interact with *cloH* (*couH*) in an adenylation reaction of L-tyrosine[128]. All three clusters contain also two positive regulators, *novE*[34] and *novG*[33] and their orthologs, as well as a $gyrB^R$ resistance gene, coding an aminocoumarin-resistant gyrase B subunit. The gene clusters of clorobiocin and coumermycin A_1 contain an additional resistance gene $parY^{R102}$.

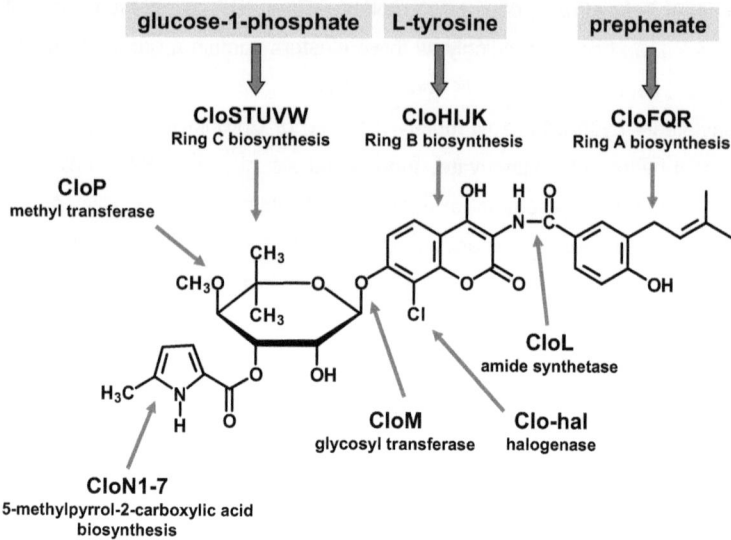

Figure I.5:
Structure of clorobiocin and function of the gene products of *cloFHIJKLMNPQRSTUVW* and *clo-hal* in clorobiocin biosynthesis[49].

This encodes an aminocoumarin resistant topoisomerase IV subunit. Clorobiocin and coumermycin A_1 are more potent inhibitors of topoisomerase IV than novobiocin, and therefore it makes sense that during the evolution of the biosynthetic gene clusters the clorobiocin and coumermycin A_1 producers had to acquire a second resistance gene.

I.2.5. Bacterial resistance mechanisms against aminocoumarins

A principal shortcoming of the aminocoumarin antibiotics is their poor activity against Gram-negative organisms which is due to a synergistic effect of the permeability barrier, imposed by the outer membrane[114], and of active efflux by multidrug efflux pumps of these organisms[81]. The outer membrane bilayer is composed of lipopolysaccharides[85]. Because of the presence of porins[12], protein complexes that cross the membrane forming a pore through which small molecules can diffuse with different selectivity, the outer membrane is permeable for small hydrophilic substances but not for hydrophobic compounds or molecules with higher molecular weight.

The multidrug efflux pump is another worrisome mechanism that contributes to bacterial antibiotic resistance. Active efflux is a mechanism responsible for the export of a variety of antibiotics outside the bacterial cell. The ability of efflux systems to recognize a large number of compounds other than their natural substrates is probably because substrate recognition is based on physiochemical properties, such as hydrophobicity, aromaticity and ionisable character, rather than on defined chemical properties as, for instance, in classical enzyme-substrate or ligand-receptor recognition. Because aminocoumarins are amphiphilic molecules they are easily recognized by many efflux pumps[84], e.g. the multidrug transporter ABC (*mdtABC* or AcrAB)[81, 7]. This efflux system usually consists of membrane protein complexes, e.g. MdtA/MdtB/MdtC or AcrA/AcrB, and a common membrane channel TolC (**Figure I.6**). Since efflux pump inhibitors, like verapamil, can be used in combination with current drugs to increase their effective intracellular concentration, the possible impact of efflux pump inhibitors together with improved uptake of the antibiotic is of great clinical interest[94].

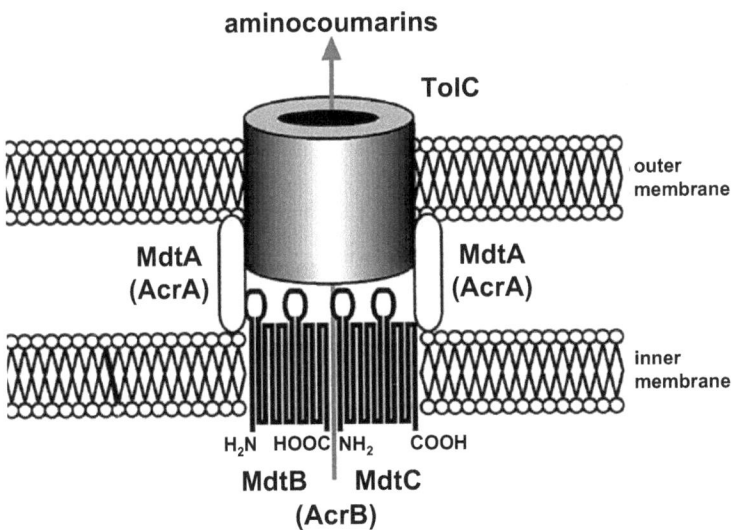

Figure I.6:
Schematic model of the molecular construction of MdtABC-TolC and AcrAB-TolC transporter complexes, respectively; adapted from ref. 81.

I.2.6. Transport of catechol siderophores

In the outer membrane bilayer reside various specific transporters for the uptake of essential nutrients. Iron is an essential trace nutrient for most known organisms for processes such as respiration and DNA synthesis. Despite being one of the most abundant elements in the earth crust, the bioavailability of iron in many oxygenated environments such as the soil or sea is limited by very low solubility of the Fe^{3+} ion, form not readily usable by organisms.

Gram-negative bacteria produce and secrete iron-chelating molecules that efficiently bind iron (siderophores) and possess specific receptors in the outer membrane (e.g. *E. coli* Cir, Fiu and FepA) that facilitate the active transport of the siderophore-iron complex into the cell[82]. Many siderophores, like *E. coli* enterobactin, contain catechol (= o-diphenol) motifs whose proximate hydroxyl groups are responsible for chelating the Fe^{3+} ion. In the case of *E. coli* Cir, Fiu and FepA transporters, they receive the energy for active transport from the inner membrane associated proteins TonB, ExbB and ExbD.

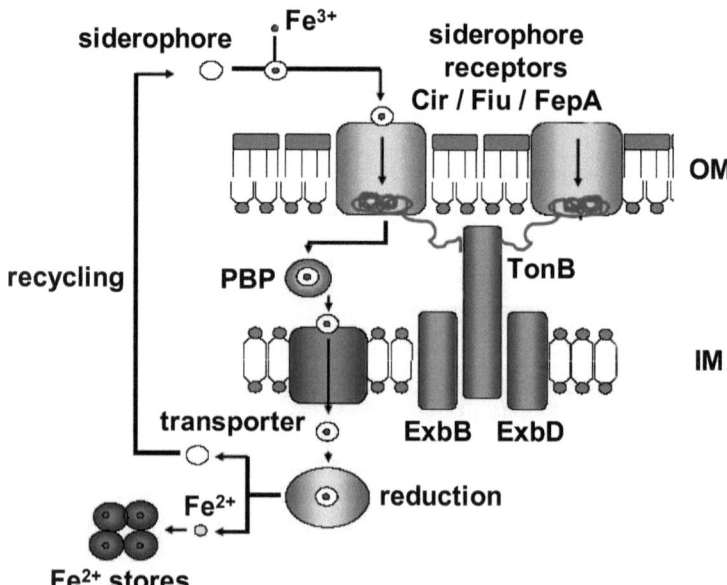

Figure I.7:
Schematic representation of the uptake of iron via siderophores in Gram-negative bacteria. OM = outer membrane; IM = inner membrane; adapted from 25.

After passing the outer membrane, the siderophore-iron(III) complex is bound by a periplasmic binding protein (PBP). The PBP donates the ferrisiderophore to a transporter and once inside the cytoplasm a reductase releases iron under the Fe^{2+} form, which can be incorporated into Fe-containing proteins. The siderophore can eventually be recycled (**Figure I.7**)[25].

The expression of this machinery is up-regulated under conditions of iron starvation. Bacterial growth during an infection in the human body represents a condition of extreme iron starvation[24], and therefore the bacterial iron uptake mechanism offers a possibility to overcome membrane-associated drug resistance by a Trojan Horse approach.

It has been shown that beta-lactam antibiotics to which a catechol moiety had been chemically attached were transported by siderophore transporters into the Gram-negative cell, resulting in an enhanced antibacterial activity[79]. As it was discussed earlier, aminocoumarin antibiotics have potent antibacterial activity against Gram-positive bacteria but little against Gram-negative microorganisms, partly due to the lack of permeability. Therefore, it was of great interest to develop aminocoumarin derivatives with siderophore-like structures, so that the aminocoumarin antibiotic is recognized and actively transported into the cell under involvement of catechol siderophore transporters (**Figure I.8**) and in this way to obtain new antibiotics that could add to the scarce set of available compounds to fight the increasing problem of multi-drug resistant Gram-negative infections[61].

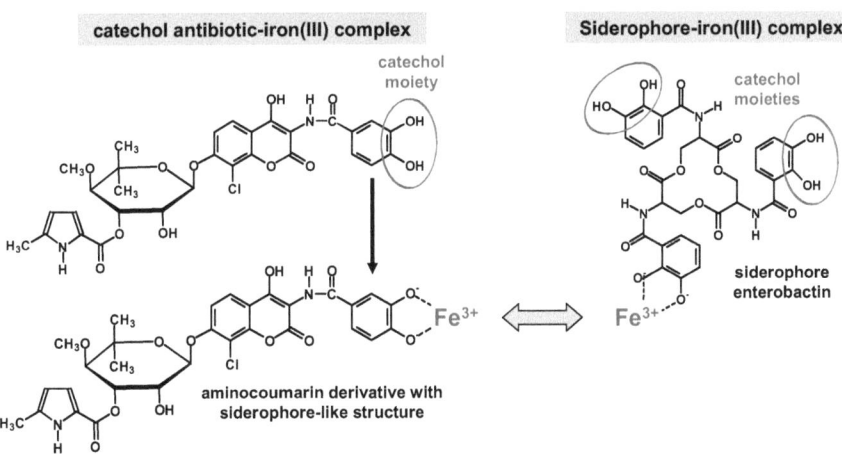

Figure I.8:
Aminocoumarin derivative imitating the siderophore-like structure (catechol) of a siderophore, e.g. enterobactin from *Escherichia coli*.

II. MATERIALS AND METHODS

II.1. Microbiology methods

II.1.1. Microorganisms

The bacterial strains used or constructed during this study are listed in **Table II.1**.

Table II.1: Microorganisms.

Strains	Relevant genotype, description, or properties	Reference
E. coli		
K-12 MG1655		17
XL1 Blue	general cloning host; recA1, endA1, gyrA96, thi-1, hsdR17, supE44, relA1, lac[F', proAB, lacI^q, ZΔM15, Tn10]; Tet^r	19 Stratagene
ET12567	DNA methylase negative strain; dam$^-$13::Tn9, dcm$^-$6, hsdM; Tet^r, Cm^r	71
BL21(DE3)/pLysS	host strain for heterologous gene expression; F$^-$, ompT, hsdS$_B$(r$_B^-$ m$_B^-$), gal, dcm, (DE3)pLysS; Cm^r	Novagen
BW25113	K-12 derivative; ΔaraBAD, ΔrhaBAD	27
JW0585-2	K-12 derivative; ΔentC; Km^r	8
JW5503-1	K-12 derivative; ΔtolC; Km^r	8
JW5195-1	K-12 derivative; ΔtonB; Km^r	8
JW3605-1	K-12 derivative; ΔrfaP; Km^r	8
SA-101	K-12 derivative; ΔtolC, ΔentC; aac(3)IV; Km^r, Apr^r	This study
SA-102	K-12 derivative; ΔtonB, ΔentC; aac(3)IV; Km^r, Apr^r	This study
SA-103	K-12 derivative; ΔtonB, ΔtolC; aac(3)IV; Km^r, Apr^r	This study
SA-104	K-12 derivative; ΔtonB, ΔtolC, ΔentC; aac(3)IV; aadA; Km^r, Apr^r, Str^r	This study
SA-105	K-12 derivative; ΔrfaP, ΔentC; aac(3)IV; Km^r, Apr^r	This study
SA-106	K-12 derivative; ΔrfaP, ΔtolC; aac(3)IV; Km^r, Apr^r	This study
SA-107	K-12 derivative; ΔrfaP, ΔtolC, ΔentC; aac(3)IV; aadA; Km^r, Apr^r, Str^r	This study
Streptomyces coelicolor		
M512	S. coelicolor M145 derivative; ΔactII-ORF4, ΔredD, SCP1$^-$, SCP2$^-$	38
M1154	S. coelicolor M145 derivative; Δact, Δred, Δcpk, Δcda, rpoB[C1298T], rpsL[A262G]	45
M512(clo-BG1)	S. coelicolor M512 containing cosmid clo-BG1; Km^r	31
M512(clo-SA2)	S. coelicolor M512 containing cosmid clo-SA2; Km^r	This study
M512(cloSA4)	S. coelicolor M512 containing cosmid clo-SA4; Km^r	This study
M1154(clo-SA2)	S. coelicolor M1154 containing cosmid clo-SA2; Km^r	This study
M512(clo-SA2)/pSA11	S. coelicolor M512 containing cosmid clo-SA2 and plasmid pSA11; Km^r, Tsr^r	This study
M1154(clo-SA2)/ pSP126$_{1110}$	S. coelicolor M1154 containing cosmid clo-SA2 and plasmid pSP126$_{1110}$; Km^r, Tsr^r	This study

II.1.2. Culture media

Unless otherwise stated, the media were prepared with distilled water and autoclaved for 20 min at 121°C. All the recipes for the media are given for 1 l of final volume. When necessary, supplementary components like antibiotics or other heat labile substances, sterilized by filtering through 0.22 µm pore sized filters, were added to the sterile media. Culture media were stored at RT or 4°C.

Culture media for E. coli

LB (Luria-Bertani) medium[70]

NaCl	10.0 g
Tryptone	10.0 g
Yeast extract	5.0 g

Components were dissolved in 1 l water, adjusted to pH 7.0 and sterilized by autoclaving.

Mueller-Hinton agar (Roth)

Mueller-Hinton agar	38.0 g

Components were dissolved in 1 l water and sterilized by autoclaving.

SOB medium

Tryptone	20.0 g
Yeast extract	5.0 g
NaCl	0.5 g

Components were dissolved in 1 l water and sterilized by autoclaving.

Culture media for Streptomyces

TSB (Tryptone Soya Broth) medium[59]

Tryptone Soya Broth	30.0 g

Components were dissolved in 1 l water and sterilized by autoclaving.

MS (Mannitol Soya flour) Agar[59]

Mannitol	20.0 g
Soya flour	20.0 g
Agar	20.0 g

The mannitol was dissolved in 1 l tap water and 100 ml each poured into flasks containing 2 g agar and 2 g soya flour. The medium was sterilized twice (115°C, 15 min) by autoclaving with moving of the media between the two runs.

GYM medium[105]

Glucose	4.0 g
yeast extract	4.0 g
Malt extract	10.0 g
Peptone	1.0 g
NaCl	2.0 g

Ingredients were dissolved in 1 l water, adjusted to pH 7.2 with NaOH and sterilized by autoclaving.

Corn starch medium (Clorobiocin preculture medium)[72]

Corn starch	10.0 g
Peptone	10.0 g
Meat extract	5.0 g

Ingredients were dissolved in 1 l water, adjusted to pH 7.0 and sterilized by autoclaving.

Distillers solubles medium (clorobiocin production medium)[72]

Distillers' solubles	48.0 g
Glucose	12.0 g
$CoCl_2 \cdot 6 H_2O$	24 mg
$CaCO_3$	6.0 g
$(NH_4)_2SO_4$ (16%)	13 ml
Glucose (25%)	100 ml

Distillers' solubles, glucose and cobalt chloride were dissolved in 887 ml water and adjusted to pH 7.8. After adding of calcium carbonate the medium was sterilized by autoclaving. After autoclaving sterile ammonium sulphate and glucose solutions were added.

CDM medium (novobiocin production medium)[60]

tri-sodium citrate $\cdot$ 2 H_2O	6.0 g
L-proline	6.0 g
$K_2HPO_4 \cdot 3 H_2O$	2.0 g
$(NH_4)_2SO_4$	1.5 g
NaCl	5.0 g
$MgSO_4 \cdot 7 H_2O$	2.05 g
$CaCl_2 \cdot 2 H_2O$	0.4 g
$FeSO_4 \cdot 7 H_2O$	0.2 g
$ZnSO_4 \cdot 7 H_2O$	0.1 g
Glucose (30%)	100 ml

The Ingredients up to NaCl were dissolved in 900 ml water and adjusted to pH 7.2. After adding of $MgSO_4$, $CaCl_2$, $FeSO_4$ and $ZnSO_4$ the medium was again adjusted to pH 7.2 and sterilized by autoclaving. The glucose solution was also sterilized by autoclaving and added afterwards to the medium.

YEME (Yest Extract Malt Extract) medium (protoplast transformation)[59]

Sucrose	340.0 g
Glucose	10.0 g
Peptone	5.0 g
Yeast extract	3.0 g
Malt extract	3.0 g
$MgCl_2$ (2.5 M)	2 ml
Glycine (20%)	25 ml

Ingredients were dissolved in 1 l water and sterilized by autoclaving. After autoclaving sterile $MgCl_2$ and glycine solutions were added.

R5 medium

Sucrose	103.0 g		
$MgCl_2 \cdot 6\ H_2O$	10.1 g		
Glucose	10.0 g		
TES	5.7 g		
Yeast extract	5.0 g		
K_2SO_4	0.25 g	Trace elements solution (for 1 l)	
Casaminoacids	0.1 g	$FeCl_3 \cdot 6\ H_2O$	200 mg
Trace elements solution	2.0 ml	$ZnCl_2$	40 mg
		$CuCl_2 \cdot 2\ H_2O$	10 mg
$CaCl_2 \cdot 2\ H_2O$ (1 M)	20 ml	$MnCl_2 \cdot 4\ H_2O$	10 mg
L-Prolin (20%)	15 ml	$Na_2B_4O_6 \cdot 10\ H_2O$	10 mg
KH_2PO_4 (0.5%)	10 ml	$(NH_4)_6Mo_7O_{24} \cdot 4\ H_2O$	10 mg

The ingredients were dissolved in 1 l water and adjusted to pH 7.2 and sterilized by autoclaving. The three supplementary solutions were also sterilized by autoclaving and added afterwards to the medium. For the preparation of agar plates 23 g agar was added after adjusting to pH 7.2. For the preparation of soft agar only 6 g agar were added. Ingredients for the trace elements solution were dissolved in 1 l water and sterilized by autoclaving.

II.1.3. Growth and preservation of microorganisms

E. coli strains

E. coli strains were routinely cultivated in liquid or on solid LB medium with appropriate antibiotics at 37°C overnight (18 h). Standard methods for cultivation were performed as described by Sambrook and co-workers[100].

Stocks of *E. coli* strains for long term storage were prepared by mixing 800 µl of overnight culture with 400 µl sterile glycerol solution (80% in distilled water) and stored at -70°C.

Streptomyces strains

Standard methods for cultivation were performed as described by Kieser and co-workers[59]. For antibiotic production in batch fermentation, *Streptomyces* strains were routinely pre-cultivated in liquid TSB medium (BD Bioscience) with appropriate antibiotics at 30°C for 2-3 days using baffled Erlenmeyer flasks with a steel spring. The cultivation was continued in distillers' solubles, GYM or CDM medium for antibiotic production at 30°C for 5-8 days.

For preparation of stocks of *Streptomyces* strains for long-term storage as frozen mycelium, 1 ml of 2-3 days old TSB culture was harvested by centrifugation, the cells were resuspended in 0.5 ml 20% glycerol and stored at -70°C.

For long term storage as spore suspensions, *Streptomyces* strains were spread on MS or R5 agar and incubated at 30°C for about 1-2 weeks until dense and matured sporulation was observed. 6 ml Tween® 20 (0.1%) were added to each plate and the spores scraped off into suspension with a sterile cotton bud. The spore suspension was poured into a falcon tube, vortexed and separated from the mycelium by passing through sterile cotton wool. After centrifugation (4000 rpm, 10 min, 4°C) spores were resuspended in 0.5 ml glycerol (20%) and stored at -70°C.

II.1.4. Antibiotic solutions

For stock solutions, antibiotics were dissolved in sterile distilled water (unless otherwise stated) and kept at –20°C. The aqueous solutions were sterilized by filtration (pore size 0.22 µm). For antibiotic selection, the required antibiotics were added to the media in appropriate concentrations.

Antibiotic	Resistance gene	Final concentration in the media (µg/ml)	Stock solution (mg/ml)
Apramycin	*acc(3)IV*	50	50
Carbenicillin	*bla*	100	100
Chloramphenicol	*cat*	25	25 (ethanol)
Kanamycin	*neo*	50	50
Streptomycin	*aadA*	10	10
Tetracycline	*tet*	12	12 (ethanol)
Thiostreptone	*tsr*	25	50 (DMSO)

II.2. Molecular biology methods

II.2.1. Vectors and constructs used in this study

Cloning and expression vectors used for this study and the constructs generated during this work are listed in **Table II.2**. Standard methods for DNA isolation and manipulation were performed as described by Sambrook and co-workers[100] and Kieser and co-workers[59].

Table II.2: Vectors and constructs.

Plasmids / Cosmids	Description and Properties	Reference / Source
Plasmids		
pJ201	pUC origin; Km^r	DNA2.0
pUWL201	E. coli-Streptomyces shuttle vector; ermE* promoter; pIJ101 origin; Amp^r, Tsr^r	28
pUG019	pBlueskript SK(-) derivative; aac(3)IV; Amp^r	31
pQE70	expression vector; C-terminal (His)$_6$-Tag; T5-promotor; ColE1 origin; Amp^r	Qiagen
pET22b	expression vector; C-terminal (His)$_6$-Tag; f1 origin; Amp^r	Novagen
pREP4	lac repressor plasmid; p15Aori; Km^r	Qiagen
pBR322	origin; Amp^r, Tet^r	100
pIJ790	λ-RED (gam, bet, exo), cat, araC, rep101^{ts}	47
pIJ778	pBlueskript SK(-) derivative; aadA; Str^r	47
pIJ773	pBlueskript SK(-) derivative; aac(3)IV, oriT; Apr^r	47
pSH2	pUWL201 derivative containing simL; XbaI-HindIII restriction sites; Amp^r, Tsr^r	47
pJJM301	pQE70 derivative containing dhbE; SphI-BamHI restriction sites; Amp^r	75
pET11cSaparC	pET11c derivative containing parC; restriction sites; pBR322 origin; Amp^r	Novagen
pET11cSaparE	pET11c derivative containing parC; restriction sites; pBR322 origin; Amp^r	Novagen
pSP126$_{1110}$	pUWL201 derivative containing dhbA, dhbC, dhbE, dhbB; EcoRV restriction sites; Amp^r, Tsr^r	89
pSA07	pET22b derivative containing parC from pET11c-SaparC; NcoI-XhoI restriction sites; Amp^r	This study
pSA08	pET22b derivative containing parC from pET11c-SaparE; NdeI-XhoI restriction sites; Amp^r	This study
pSA09	pQE70 derivative containing parC from pET11c-SaparC; SphI-BglI restriction sites; Amp^r	This study

pSA10	pQE70 derivative containing *parE* from pET11c-Sa*parE*; SphI-BgII restriction sites; Amp^r	This study
pSA11	pUWL201derivative containing a synthetic *ubiC/pobA* DNA fragment; HindIII-SpeI restriction sites; Amp^r, Tsr^r	This study
Cosmids		
clo-BG1	SuperCos1 derivative containing the clorobiocin biosynthetic gene cluster; *oriT, tet, attP, int* ΦC31; Km^r	31
clo-SA1	clo-BG1 derivative; *cloQ* replaced by *aac(3)IV* resistance cassette; Apr^r, Km^r	This study
clo-SA2	clo-BG1 derivative; Δ*cloQ*; Km^r	This study
clo-SA3	clo-BG1 derivative; *cloHIJK* replaced by *aac(3)IV* resistance cassette; Apr^r, Km^r	This study
clo-SA4	clo-BG1 derivative; Δ*cloHIJK*; Km^r	This study

II.2.2. DNA isolation

Unless otherwise described, buffers and solutions were prepared with distilled water, autoclaved, and stored at room temperature.

Alkaline lysis for plasmid and cosmid isolation from E. coli

Miniprep solution 1

Tris-HCl	50 mM
EDTA	10 mM
RNase A	100 µg/ml

Miniprep solution 2

NaOH	0.2 M
SDS	1%

Miniprep solution 3

Potassium acetate·pH 5.5	3 M

Tris-HCl and EDTA of solution 1 were dissolved in water and adjusted to pH 8.0. After autoclaving RNase A was added. Solution 3 was adjusted to pH 4.8 and stored at 4°C.

Alkaline-lysis method was used to isolate recombinant plasmids from *E. coli*. 2 ml of culture were centrifuged (13000 rpm, 5 min, 4°C) and the pellet resuspended in 250 µl miniprep solution 1 by vortexing. 250 µl miniprep solution II were added, mixed by inversion and 5 min incubated at room temperature. The suspension was mixed with 250 µl miniprep solution 3, incubated for 5 min on ice and centrifuged (13000 rpm, 10 min, 4°C). The supernatant was transferred into a fresh tube, mixed with 400 µl Rotiphenol® (Roth) by vortexing and centrifuged (13000 rpm, 5 min, 4°C). The supernatant was transferred into a fresh tube, 460 µl isopropanol added and mixed by inversion, and the DNA precipitated by centrifugation (13000 rpm, 30 min, 4°C). The pellet was washed with 500 µl ethanol (70%), air dried and resuspended in 20 µl Tris-HCl buffer (10 mM).

Eckhard-Lysis for fast plasmid isolation and test from E. coli colonies

Eckhardt gel

TAE buffer (1x) (see II.2.5)	125 ml
SDS	0.25 g
Agarose	1.25 g

Eckhardt buffer

Sucrose	6.25 g
Ficoll 400	0.75 g
TEA buffer (1x)	25 ml
Bromphenol blue	0.01%

To 1 ml of Eckhardt buffer 1 µl RNase A (100 mg/ml) and 5 mg lysozyme were added.

Single colonies were picked with toothpicks and rubbed into 17 µl Eckhardt buffer (microplates are suitable). Samples were put directly on the Eckhardt gel and run for the first 15 min with 20 V and after that with 90 V. Before staining with ethidium bromide, the SDS was removed by washing the gel with distilled water.

Plasmid isolation from Streptomyces

Miniprep solution 1

Glucose	50 mM
Tris-HCl	25 mM
EDTA	10 mM
RNase A	100 µg/ml
Lysozyme	4 mg/ml

Tris-HCl and EDTA of solution 1 were dissolved in water and adjusted to pH 8.0. After autoclaving RNase A and lysozyme were added.

Miniprep solution 2

NaOH	0.2 M
SDS	1%

Miniprep solution 3
Potassium acetate   5 M

Solution 3 was adjusted to pH 4.8 and stored at 4°C.

Alkaline-lysis was used to isolate recombinant plasmids from *Streptomyces*. For isolation of plasmids from *Streptomyces* 500 µl miniprep solution 1 for *Streptomyces* was added after the resuspension of the pellet in 1 ml miniprep solution 1 for *E. coli*. The suspension was incubated at 37°C for 60 min with inversion of the tubes every 10 min. The procedure was continued as described under "Alkaline lysis for plasmid and cosmid isolation from *E. coli*"

II.2.3. DNA quantification and manipulation with enzymes

DNA quantification
DNA was quantified by comparing the fluorescence intensity with the GeneRuler^{TM} 1 kb DNA ladder (Fermentas) on an agarose gel after staining with ethidium bromide.

Restriction digest
Restriction of DNA with endonucleases was performed according to the instructions provided by manufacturers (Amersham; New England Biolabs).

Ligation
DNA ligation was performed by using 1U T4-DNA ligase (Amersham), 1x ligation buffer and the mixture of insert and linearized vector at a 1:1 ratio in a total volume of 10 µl. The mixture was incubated at 4°C over night.

Dephosphorylation
This reaction avoids intramolecular ligation by dephosphorylation of the 5'-end of DNA. 0.2 U alkine phosphatase (Amersham), 1x alkaline phosphatase buffer and DNA were incubated for 1 h at 37°C.

II.2.4. PCR amplification of DNA

PCR amplification was performed with an iCycler® PCR-System (Bio-Rad). The oligonucleotide primers, listed in **Table II.3**, were purchased to MWG-Biotech. PCR amplifications for plasmid construction and gene inactivation by PCR targeting[47] were performed with the Expand High Fidelity PCR System (Roche) and the Phusion Polymerase (New England BioLabs). Colony PCR was performed with Taq-DNA polymerase (New England BioLabs).

PCR products were purified using a High Pure PCR Product Kit (Roche) and 2 μl analysed on an agarose gel as described in section II.2.5.

Table II.3: Oligonucleotides.

Name	Sequence (5'-3')	Restriction sites
Primers for cloQ inactivation		
cloQ_f	GGC GCG CCC ATT GCT CAC CGT CTT ACC GAC ACC GTC CTT ATT CCG GGG ATC TCT AGA TC	XbaI
cloQ_r	TCC CAT GGT CGA TTC CGT GTG TTG GTG AAG TGC GCG CAG ACT AGT CTG GAG CTG CTT C	SpeI
Primers for cloHIJK inactivation (resulting in cosmid clo-SA3 / clo-SA4)		
cloHIJK_f	GTA TGT TCC AAT GGC ATG GAG ACT TAA GGG GGA AGT TTG TCT AGA ATT CCG GGG ATC CGT CGA CC	XbaI
cloHIJK_r	GTG CTC CGG TCC GTG GTC CTT GTT CGC CAC CAG TGA CTA ACT AGT TGT AGG CTG GAG CTG CTT C	SpeI
Primers for construction of plasmid pSA07		
parC_f_NcoI	C ATG CCA TGG TGA GTG AAA TAA TTC AAG ATT TAT CA	NcoI
parC_f_XhoI	CCG CTC GAG GCT AAT ATA CAT GTC TAT TAC TTC AC	XhoI
Primers for construction of plasmid pSA08		
parE_f_NdeI	GG GAA TTC CAT ATG AAT AAA CAA AAT AAT TAT TCA GAT GAT TCA ATA	NdeI
parE_f_XhoI	CCG CTC GAG GAT TTC CTC CTC ATC AAA TTG A	XhoI
Primers for construction of plasmid pSA09		
parC_f	A AAG GCA TGC ATA GTG AAA TAA TTC AAG ATT TAT CAC TT	SphI
parC_r	GGA AGA TCT GCT AAT ATA CAT GTC TAT TAC TTC AC	BglII

Primers for construction of plasmid pSA10

parE_f	A AAG <u>GCA TGC</u> ATA ATA AAC AAA ATA ATT ATT CAG ATG ATT CA ATA	SphI
parE_r	GGA <u>AGA TCT</u> GAT TTC CTC CTC ATC AAA TTG ATC	BglII

Primers for construction of the E. coli mutant strains

entC_f	T CAT TAT TAA AGC CTT TAT CAT TTT GTG GAG GAT GAT **ATG** att ccg ggg atc cgt cga cc
entC_r	C CGG CCA GCG GGT GAA TGG AAT GCT CAT CCT CGC TCC **TTA** tgt agg ctg gag ctg ctt c
tolC_f	G ATC GCG CTA AAT ACT GCT TCA CCA CAA GGA ATG CAA **ATG** att ccg ggg atc cgt cga cc
tolC_r	A CGT TCA GAC GGG GCC GAA GCC CCG TCG TCG TCA **TCA** tgt agg ctg gag ctg ctt c

Primers for testing of the E. coli mutant strains by colony PCR

entC_T1	GAG TTG CAG ATT GCG TTA CC
entC_T2	CGT CAG AAT GTC GGT CAG CG
tolC_T1	CAT TAA CGC CCT ATG GCA CG
tolC_T2	GAA TAG AGG ATG GCT GGT CG
tonB_T1	TGT CTT TGT TAA GGC CAT GC
tonB_T2	TTG GGC AAC GCT ATA AAG CG
rfaP_T1	GCC CCA GCC ATG CAT TAT CC
rfaP_T2	AGT CGC CAT TGC GAA TGG CC

II.2.5. Agarose gel electrophoresis of DNA

<u>50x TAE buffer</u>

Tris base	2 M
EDTA	0.05 M
Glacial acetic acid	57.1 ml/l

The pH was adjusted to 8.0 with glacial acetic acid.

<u>Loading buffer</u>

Glycerol	49.75%
Tap water	50%
Bromphenol blue	0.25%

<u>Ethidium bromide solution</u>

Ethidium bromide	1 mg/l

1% agarosese gels (Biozym) were prepared with 1x TAE buffer to separate DNA fragments between 0.25 and 10 kb. As marker, the GeneRuler™ 1 kb DNA ladder (Fermentas) was used. Gels were run in 1x TAE buffer, stained with ethidium bromide solution for 15 min and analysed with UV light (312 nm) with an Eagle Eye II System (Stratagene) or a gel documentation system from Biostep equipped with a Argus X1 software.

DNA fragments were isolated from agarose gels using a NucleoSpin® 2 in 1 extraction kit (Macherey-Nagel) according to the protocol supplied by the manufacturer.

II.2.6. Introduction of DNA in *E. coli* and *Streptomyces*

Electroporation of E. coli cells

Preparation of electro-competent *E. coli* cells:
100 ml LB medium was inoculated with 1 ml *E. coli* overnight culture and cultivated at 37°C until the OD_{600} reached 0.6. The cells were harvested by centrifugation (5000 rpm, 10 min, 4°C) and washed twice with 30 ml ice-cold glycerol solution (10%). The pellet was suspended in the remains of the discarded supernatant. The competent cells were used immediately or stored in 50 µl aliquots at -70°C.

Electroporation:
3 µl DNA (about 100 ng) were mixed with 50 µl competent cells. The mixture was transferred into an ice-cold electroporation cuvette (0.2 cm) and electroporation was carried out with 2.5 kV using an electroporator (BioRad). The optimal time constant is 4.5 – 5.0 ms. 500 µl cold LB medium was immediately added to the electroporated cells. After 1 h incubation of the cells at 37°C, 200 µl were spread on LB agar containing an appropriate antibiotics. The agar plates were incubated overnight at 37°C.

PEG-mediated protoplast transformation of Streptomyces

The following sterile solutions were prepared separately before mixing the buffers. The buffers were stored at -20°C.

	Stock solution	Amount
P(protoplast)-buffer		
Sucrose	12%	85.5 ml
TES	0.25 M, pH 7.2	10.0 ml
$MgCl_2 \cdot 6 H_2O$	1 M	1.0 ml
K_2SO_4	140 mM	1.0 ml
KH_2PO_4	40 mM	1.0 ml
$CaCl_2 \cdot 2 H_2O$	250 mM	1.0 ml
Trace element solution	(see R5 medium)	0.2 ml
distilled water		to 100 ml
T(transformation)-buffer		
PEG 1000	50%	5.0 ml
Sucrose	25%	1.0 ml
$CaCl_2 \cdot 2 H_2O$	5 M	1.0 ml
Tris-maleate	0.5 M, pH 8.0	1.0 ml
K_2SO_4	140 mM	0.1 ml
KH_2PO_4	40 mM	0.1 ml
$MgCl_2 \cdot 6 H_2O$	1 M	0.1 ml
Trace element solution	(see R5 medium)	0.03 ml
distilled water		to 10 ml

Protoplast preparation:

20 µl spores were added to 1 ml TSB medium and incubated for 5 min at 50°C. After 2 h incubation at 30°C, 50 ml YEME medium with $MgCl_2$ and glycin were inoculated and cultivated for 2 d. The mycelium was harvested by centrifugation (5000 rpm, 10 min, 4°C) and washed twice with 20 ml sucrose solution (10.3%). The pellet was resuspended in sterile filtered lysozyme solution (4 ml P-buffer for 1 g cell pellet) and incubated for 30 – 60 min at 30°C with gentle moving. Protoplast formation was identified by microscopy and the reaction was stopped by incubation on ice. 10 ml ice-cold P buffer were added, gently mixed with the protoplasts by pipetting and filtered through cotton wool. After centrifugation (3000 rpm, 10 min, 4°C) the supernatant was discarded and the pellet carefully resuspended in the remaining liquid. Protoplasts were used immediately for transformation or stored in 100 µl aliquots at -70°C.

Protoplast transformation:

Before transformation, plasmid DNA was isolated from the DNA-methylation deficient strain *E. coli* ET12567. 10 µl (about 15 µg) plasmid DNA were added to 100 µl protoplasts and 100 µl P buffer. The suspension was mixed carefully by pipetting. The tube was inverted several times after adding of 500 µl T buffer and incubated for 1 min at room temperature. 100 µl, 200 µl and 400 µl of this suspension were mixed with 3 ml melted R5 soft agar each and poured on R5 agar plates. After 24 h incubation at 30°C the plates were overlaid with 3 ml R5 soft agar including the required antibiotics for selection. The incubation was continued for 7 d.

II.2.7 Construction and heterologous expression of the plasmid pSA11

Design of the synthetic gene operon for 3,4-DHBA biosynthesis
The nucleotide sequences of the genes *pobA* coding for a 4-hydroxybenzoate hydroxylase from *Corynebacterium cyclohexanicum* (GenBank AB210281) and *ubiC* coding for a chorismate pyruvate-lyase from *Escherichia coli* 536 (GenBank NC_008253) were redesigned using the codon preference of *S. coelicolor* M512 (DNA2.0 Gene Designer Software). The two genes were linked by translational coupling and flanked by *Spe*I and *Hind*III restriction sites. The DNA fragment was synthesized by DNA2.0 company (California, USA) and provided in the vector pJ201. This nucleotide sequence of the synthetic *ubiC/pobA* DNA fragment has been listed in **Figure III.4**.

Cloning of the synthetic gene operon in an expression vector
The synthetic *ubiC/pobA* fragment was isolated from the pJ201 vector (DNA2.0) by restriction digest with *Spe*I and *Hind*III and cloned into pUWL201 which contains the constitutive *ermE** promoter for foreign gene expression. Transformation of pSA11 into the integration mutant *S. coelicolor* M512(clo-SA2) was carried out by PEG-mediated protoplast transformation (II.2.6). Transformed colonies appeared after 5 d at 30 °C.

II.2.8 Construction and heterologous expression of cosmid clo-SA2

In cosmid clo-BG1, *cloQ* was replaced by RED/ET-mediated recombination with an apramycin-resistance (*aac(3)IV*) cassette that was flanked by *Xba*I and *Spe*I recognition sites. For replacement of *cloQ*, the cassette was generated by PCR using pUG019 as template and the primers cloQ_f and cloQ_r (**Table II.3**). PCR amplification was performed in 50 µl volume with 100 ng template, 0.25 mM dNTPs, 50 pmol of each primer, and 5% (v/v) DMSO with the Expand High Fidelity PCR system (Roche): denaturation at 94 °C for

2 min, then 10 cycles with denaturation at 94 °C for 45 s, annealing at 50 °C for 45 s, and elongation at 72 °C for 90 s, followed by 15 cycles with annealing at 55 °C for 45 s, and the last elongation step at 72 °C for 5 min. The PCR product was introduced by electroporation into *E. coli* BW25113/pIJ790 harboring cosmid clo-BG1. The resulting modified cosmid clo-SA1 was isolated, transformed into the nonmethylating strain *E. coli* ET12567, reisolated, and digested with *Xba*I and *Spe*I to remove the apramycin-resistance cassette. Religation overnight at 4 °C gave the cosmid clo-SA2.

The cosmid clo-SA2 isolated from *E. coli* ET12567 was introduced into *S. coelicolor* M512 by PEG-mediated protoplast transformation. Clones resistant to kanamycin were selected. Feeding experiments were carried out by addition of 1 mg Ring A (3-dimethylallyl-4-hydroxybenzoic acid) dissolved in 100 μl ethanol to 80 ml *S. coelicolor* M512(clo-SA2) culture in distillers` solubles medium.

For mutasynthesis experiments, 3 mg of the respective catechol compounds were dissolved in 100 μl ethanol and added to 80 ml of the culture of the *cloQ* defective strain *S. coelicolor* M512(clo-SA2) in distillers` solubles medium one day after inoculation. After 5-8 days cultivation at 30 °C and 210 rpm, the cultures were extracted and analyzed by HPLC described below.

II.2.9. Construction and heterologous expression of cosmid clo-SA4

In cosmid clo-BG1 *cloHIJK* was replaced by RED/ET-mediated recombination with an apramycin-resistance (*aac(3)IV*) cassette that was flanked by *Xba*I and *Spe*I recognition sites. For replacement of *cloHIJK*, the cassette was generated by PCR by using pIJ773 as template and the primers *cloHIJK*_f and *cloHIJK*_r (**Table II.3**). PCR amplification was performed in 50 μl volume with 100 ng template, 0.25 mM dNTPs, 50 pmol of each primer, and 5% (v/v) DMSO with the Expand High Fidelity PCR system (Roche): denaturation at 94 °C for 2 min, then 10 cycles with denaturation at 94 °C for 45 s, annealing at 50 °C for 45 s, and elongation at 72 °C for 90 s, followed by 15 cycles with annealing at 55 °C for 45 s, and the last elongation step at 72 °C for 5 min. The PCR product was introduced by electroporation into *E. coli* BW25113/pIJ790 harboring cosmid clo-BG1. The resulting modified cosmid clo-SA3 was isolated, transformed into the nonmethylating strain *E. coli* ET12567, reisolated, and digested with *Xba*I and *Spe*I to remove the apramycin-resistance cassette. Religation overnight at 4 °C gave the cosmid clo-SA4.

The cosmid clo-SA4 isolated from *E. coli* ET12567 was introduced into *S. coelicolor* M512 by PEG-mediated protoplast transformation. Clones resistant to kanamycin were selected. Feeding experiments were carried out by addition of 2 mg Ring B (3-amino-4,7-dihydroxy-8-methyl-coumarin) dissolved in 100 µl ethanol to 80 ml *S. coelicolor* M512(clo-SA4) culture in distillers' solubles medium. After 5-8 days cultivation at 30 °C and 210 rpm, the cultures were extracted and analyzed by HPLC as described under II.4.1.

II.2.10. Generation of *E. coli* mutants

E. coli double and triple mutants were generated from *E. coli* K-12 MG1655 mutants obtained from the Keio collection. The gene *entC* was replaced in *E. coli* JW5195-1/pIJ790 and JW5503-1/pIJ790 using RED/ET-mediated recombination. An apramycin resistance cassette (*acc(3)IV*) was amplified from plasmid pIJ773. The primers *entC*_f and *entC*_r used for PCR amplification are listed in **Table II.3**. The gene *tolC* was replaced in *E. coli* JW5195-1/pIJ790 using RED/ET-mediated recombination. For this purpose, a streptomycin resistance cassette (*aadA*) was amplified from plasmid pIJ778 using the primer pair *tolC*_f and *tolC*_r (**Table II.3**). The genotype of the resulting mutants was confirmed by PCR with chromosomal DNA.

II.3. Biochemistry methods

II.3.1. Assay compounds, enzymes, DNAs and chemicals

Reference compounds

Novobiocic acid	Isolated from *S. spheroides* AM1T2
Novobiocin	Sigma-Aldrich
Clorobiocin	A. Maxwell (John Innes Centre Norwich, UK)
Coumermycin A_1	Sigma-Aldrich
Simocyclinone D8	H.-P. Fiedler (University of Tübingen, Germany).
Ring B (3-amino-4,7-dihydroxy-8-methyl-coumarin)	Pharmacia & Upjohn, Inc. (Kalamazoo, MI)
Ring A (3-Dimethylallyl-4-hydroxybenzoic acid)	Obtained by hydrolysis of novobiocin
Catechol compounds	Sigma-Aldrich

Enterobactin	Sigma-Aldrich

Enzymes

E. coli DNA gyrase	Inspiralis (Norwich, UK)
E. coli DNA topoisomerase IV	Inspiralis (Norwich, UK)
S. aureus DNA gyrase	Inspiralis (Norwich, UK)

DNAs

relaxed pBR322 DNA	Inspiralis (Norwich, UK)
kDNA (from Crithidia fasciculate)	Inspiralis (Norwich, UK)

Chemicals

Potassium glutamate	Fluka
Sodium glutamate	Sigma-Aldrich
KCl	Sigma-Aldrich
NaCl	Sigma-Aldrich
2,2'-bipyridyl	Sigma-Aldrich
2,3,5-triphenyltetrazolium chloride	Roth

II.3.2. General methods for protein expression and purification

Overexpression and purification of His_6-Tag proteins from E. coli were carried out as described by the user manual of Qiagen "A handbook for high level expression and purification of 6x His-tagged proteins". The buffers were prepared with distilled water, autoclaved, filtered through 0.2 µm pore sized filters and stored at 4°C. If required, imidazol, β-mercaptoethanol, lysozyme, PMSF and DTT were added fresh before use.

<u>Lysis buffer</u>

Tris-HCl (pH 8.0)	50 mM
NaCl	500 mM
Glycerol	10%
Tween® 20	1%
Imidazol	20 mM
β-Mercaptoethanol	10 mM
Lysozyme	0.5 mg/ml
PMSF	0.5 mM

<u>Washing buffer</u>

Tris-HCl (pH 8.0)	50 mM
NaCl	500 mM

Glycerol	10%
Imidazol	20 mM
β-Mercaptoethanol	10 mM

Elution buffer

Tris-HCl (pH 8.0)	50 mM
NaCl	500 mM
Glycerol	10%
Imidazol	250 mM
β-Mercaptoethanol	10 mM

Dialysis buffer

Tris-HCl (pH 8.0)	25 mM
NaCl	100 mM
DTT	2 mM
Glycerol	15%

II.3.3. Cloning, protein expression, and purification of S. aureus topoisomerase IV subunits ParC and ParE

S. aureus RN4220 parC and parE (GenBank D67075) were PCR-amplified using plasmids pET11c-SaparC and pET11c-SaparE kindly provided by H. Hiasa (University of Minnesota) as a template. The primers ParC_f/ParC_r and ParE_f/ParE_r are listed in Table II.3. parC and parE were cloned into the SphI and BglII restriction sites of pQE70 expression vector and transformed in E. coli BL21/pREP4 cells. For protein expression, cultures were grown at 37°C in 1 l of LB medium to OD_{600} = 0.7. 1 mM IPTG was added and growth continued for 4 h at 37°C. After centrifugation cell pellets were resuspended in 25 ml lysis buffer and sonicated to release soluble proteins. Insoluble material was removed by centrifugation and the supernatant loaded onto a 5 ml His-Trap™ HP column (GE Healthcare) that had been equilibrated previously with washing buffer. The column was eluted with a linear gradient from 20 – 250 mM imidazole (30 min; 1 ml/min) and the eluates were dialyzed against the dialysis buffer. The final yields were 7 mg/l of ParC and 5 mg/l of ParE, respectively. Absorbance at 280 nm was measured to calculate protein concentration. SDS polyacrylamide gel electrophoresis of the purified topoisomerase IV subunits revealed bands with apparent molecular weights of 96.2 kDa and 77.8 kDa, corresponding to ParC and ParE, respectively. S. aureus topoisomerases IV were reconstituted *in vitro* by mixing equimolar amounts of the subunits and incubation on ice for 10 min.

II.3.4. Cloning, protein expression, and purification of the AMP ligase DhbE

The *E. coli* strain BL21(DE3)pLysS/pREP4 with the integrated plasmid pJJM301 was kindly provided by M. Marahiel (University of Marburg, Germany). For protein expression, cultures were grown at 37°C in 1 l of LB medium to OD_{600} = 0.7. 500 µM IPTG was added and growth continued for 3 h at 30°C. After centrifugation cell pellets were resuspended in 12.5 ml lysis buffer and sonicated to release soluble proteins. Insoluble material was removed by centrifugation and the supernatant loaded onto a 5 ml His-Trap^{TM} HP column (GE Healthcare) that had been equilibrated previously with washing buffer. The column was eluted twice with elution buffer and the eluates were dialyzed against the dialysis buffer. The final yields were 21 mg of purified protein DhbE of one litre culture. Absorbance at 280 nm was measured to calculate protein concentration. SDS polyacrylamide gel electrophoresis of the purified protein revealed a band with the apparent molecular weight of 59.9 kDa, corresponding to DhbE. DhbE was used in amide synthetase assays with the catechol substrat 2,3-dihydroxybenzoic acid as described under II.3.6.

II.3.5. Denaturing polyacrylamide gel electrophoresis (SDS-PAGE) and Coomassie staining

	Stock solution	Amount
1 Stacking gel (4%)		
Distilled water		1192 µl
Tris-HCl (pH 6.8)	0.5 M, pH 6.8	500 µl
SDS	10%	20 µl
Rotiphorese® Gel 30	30%	266 µl
APS	10%	20 µl
TEMED	100%	2 µl
1 Running gel (12%)		
Distilled water		1848 µl
Tris-HCl	1.5 M, pH 8.8	1000 µl

SDS	10%	40 µl
Rotiphorese® Gel 30	30%	1068 µl
APS	10%	40 µl
TEMED	100%	4 µl

Laemmli buffer (4x)

Tris-HCl	1 M, pH 6.8	2.5 ml
Glycerol	100%	4 ml
SDS		0.8 g
Bromphenol blue		0.4 mg
ß-Mercaptoethanol	100%	2.0 ml
Distilled water		ad 10 ml

Electrophoresis buffer (10x)

Tris base	30.3 g
Gycine	144 g
SDS	10.0 g
Distilled water	ad 1 l

Fixing buffer

Distilled water	70%
Acetic acid	10%
Methanol	20%

Staining buffer

Coomassie Brilliant Blue G-250	0.25%
Distilled water	45%
Acetic acid	10%
Methanol	45%

Bleaching buffer

Distilled water	45%
Acetic acid	10%
Methanol	45%

The SDS-PAGE was carried out according to the method of Laemmli[62]. Sample and sample buffer were mixed in the ratio 3:1 (total volume 40 µl) and incubated for 3 min in boiling water (95°C). 4% and 12% polyacrylamide gels were used as stacking and running gels respectively. Gel electrophoresis was carried out at 100 V during the run in the stacking gel and later at 200 V using a Mini-PROTEAN II electrophoresis cell (BioRad). Proteins were stained with Coomassie Brilliant Blue G-250 solution for 15 min after treatment of the gels with fixing buffer for 5 min. The following bleaching of the gels was carried out with bleaching solution for 2 h. To determine protein sizes, 2 µl of protein marker (LMW Calibration Kit for SDS Electrophoresis, GE Healthcare) were used.

II.3.6. Amide synthetase assay

The assay contained the listed components in a final volume of 100 µl. The acyl substrates are 3,4-dihydroxybenzoic acid, 2,3-dihydroxybenzoic acid, caffeic acid, 3,4-dihydroxyphenylacetic acid, 3,4-dihydroxypropionic acid or 3,4-dihydroxymandelic acid, respectively.

Acyl substrate	2 mM
Ring B	2 mM
ATP	5 mM
Mg_2Cl	5 mM
Ascorbic acid	10 mM
Tris-HCl (pH 8.0)	100 mM
Acyl ligase	10 µg

The reaction was carried out for 45 min at 30°C and stopped by addition of 5 µl 1.5 M trichloroacetic acid. The reaction mixture was extracted with 100 µl ethyl acetate. The organic layer was used for analysis. After evaporation of the solvent and dissolution in 100 µl methanol, the sample was analyzed by HPLC. A linear gradient from 60% to 100% solvent B (solvent A H_2O/HCOOH 99:1; solvent B MeOH/HCOOH 99:1) over 30 min was used. UV detection was carried out at 330 nm. Novobiocic acid was used as standard.

II.3.7. Topoisomerase IV decatenation assay

New aminocoumarin compounds were obtained by metabolic engineering, mutasynthesis and chemoenzymatic synthesis as described previously. These compounds were dissolved in a small volume of dimethyl sulfoxide (DMSO) and the solutions were diluted with water to a final concentration of 5% DMSO. Topoisomerase IV activity was measured by using a decatenation assay that monitored the ATP-dependent unlinking of DNA minicircles from kDNA.

Assay buffer

HEPES-KOH (pH 7.5)	40 mM
Potassium glutamate	100 mM
Magnesium acetate	10 mM
DTT	10 mM
ATP	1.2 mM
Albumin	50 µg/ml

Dilution buffer

HEPES-KOH (pH 7.5)	40 mM
Potassium glutamate	100 mM
EDTA	1 mM
DTT	1 mM
Glycerol	40%

Stop buffer

Sucrose	40%
Tris-HCl (pH 7.5)	100 mM
EDTA	100 mM
Bromphenol blue	0.25%

Topoisomerase IV decatenation assays (30 µl) were performed by incubating 200 ng of *S. aureus* topoisomerase IV or 2 U of *E. coli* topoisomerase IV and 200 ng kDNA with the indicated concentrations of the aminocoumarins in the assay buffer. The final DMSO concentration in the assay did not exceed 0.5% (v/v). The reactions were performed at 37°C and terminated after 45 min by addition of an equal volume of stop buffer, followed by extraction with one volume of chloroform/iso-amyl alcohol (24:1). The aqueous phase was analyzed on 1% agarose gels for 4 h at 80 V in TAE and visualized after staining with ethidium bromide. The IC_{50} for inhibition of decatenation can be visually assessed as the concentration of compound which leads to a 50% reduction in the mini-circle band. IC_{50} values are averages from at least two separate experiments.

II.3.8. DNA gyrase supercoiling assay

The same compounds as in the topoisomerase IV decatenation assays were tested. DNA gyrase activity was measured by a supercoiling assay that monitored the ATP-dependent conversion of relaxed pBR322 DNA to the supercoiled form.

Assay buffer (5x)

Potassium glutamate	3500 mM
Tris-HCl (pH 7.5)	175 mM
KCl	120 mM
Mg_2Cl	20 mM
DTT	10 mM
Spermidin	9 mM
ATP	6 mM
Albumin	0.5 mg/ml
Glycerol	32.5%

Dilution buffer

Tris-HCl (pH 7.5)	50 mM
KCl	100 mM
DTT	2 mM
EDTA	1 mM
Glycerol	50%

Assays (30 µl) were prepared by mixing 6 µl assay buffer (5x) with 0.5 µg of the relaxed pBR322 DNA, water and either 1 U of *S. aureus* or 1 U of *E. coli* DNA gyrase dissolved in 6 µl of dilution buffer (5x). Reaction mixtures were incubated at 37°C for 45 min and terminated by adding 30 µl stop buffer (see II.3.7), followed by extraction with one volume (30 µl) chloroform/iso-amyl alcohol (24:1). Prior to agarose gel analysis, samples were subjected to a buffer exchange with 10 mM Tris-HCl (pH 8.0), performed by dialysis with MF^{TM}-membrane filters (Millipore, 0.025 µm VSWP). For this purpose, membrane filters were cut into pieces and placed on the surface of the buffer solution in a petri dish. The aqueous phase of the assay mixtures was pipetted onto the floating membranes, and the petri dish was covered with a lid to avoid evaporation. After 3 h, the assay mixtures were removed from the membranes and mixed with 15 µl loading buffer (50% water, 49.75% glycerol, and 0.25% bromophenol blue). 20 µl of each sample were analysed by agarose gel electrophoresis as described under II.2.5. The IC_{50} for inhibition of the supercoiling can be visually assessed as the concentration of compound which leads to a 50% reduction in the supercoiling band. IC_{50} values are averages from at least two separate experiments.

II.3.9. Agar diffusion tests

Agar plates (40 ml medium) were prepared by adding 1 ml of a culture of the respective *E. coli* mutant (overnight culture in LB medium, OD_{600} 1.2) (**Table II.1**) and 50 µM 2,2´-bipyridyl in melted Mueller-Hinton agar (Roth). Different amounts of antibiotic, dissolved in

15 µl methanol, were applied to paper disks of 7 mm of diameter. These disks were placed on the agar and the plates were incubated at 37 °C for 16 h. For visualization of living cells, plates were flooded with 5 ml of 0.5% aqueous 2,3,5-triphenyltetrazolium chloride (Roth), and after incubation for 10 min at 37°C the inhibition zones were determined. The minimal inhibitory concentration (MIC) was determined according to Wiegand and co-workers[123].

II.4. Analytical chemistry techniques

II.4.1. Production and purification of novclobiocin 401

S. coelicolor M512(clo-SA2)/pSA11 was precultivated in TSB medium (BD Bioscience) with 50 µg/ml thiostrepton for 3 days and then used to inoculate 40 flasks, each containing 50 ml CDM production medium with 50 µg/ml thiostrepton. The cultivation was carried out for 5 days at 30 °C and 210 rpm. The culture was adjusted to pH 4 with hydrochloric acid and extracted twice with an equal volume of ethyl acetate. The organic layer was evaporated to dryness. The residue was dissolved in methanol and purified by preparative HPLC with a linear gradient from 70% to 100% solvent B in solvent A over 36 min (solvent A H_2O/HCOOH 99:1; solvent B MeOH: HCOOH 99:1).

For analytic purposes, 1 ml bacterial culture was acidified with HCl to pH 4 and extracted with equal volume of ethyl acetate. After evaporation of the solvent, the residue was re-dissolved in 100 µl methanol. 80 µl were analyzed by HPLC with a linear gradient from 60% to 100% solvent B over 23 min. UV detection was carried out at 280 nm.

II.4.2. HPLC analysis

For HPLC analysis an Agilent 1100 system with ChemStation software was used. For analytical purpose, samples were analysed in a Multosphere RP18-5 column (250 x 4 mm, 5 µm; Agilent) at a flow rate of 1 ml/min. For preparative HPLC analysis extracts were purified using a Multosphere column 120 RP18-5 (250 x 20 mm, 5 µm; C&S Chromatographie Service Düren, Germany) at a flow rate of 2 ml/min. For detailed information see II.4.1.

II.4.3. LC-MS analysis

For LC-ESI-MS analysis an electrospray ionization (ESI) mass spectrometer (LC/MSD Ultra Trap System XCT 6330; Agilent Technology) was used. Culture extracts were prepared as described under II.4.1. Samples were analysed on a Nucleosil 100-C18 column (100 x 2 mm, 3 µm) at a temperature of 40 °C and a flow rate of 400 µl/min. A linear gradient of solvent A (H_2O/HCOOH 99.9:0.1) and solvent B (methanol/HCOOH 99.94:0.06) from 60% to 100% of solvent B over 20 min was used. UV detection was carried out at 280 nm and authentic clorobiocin was used as a standard. ESI-MS detection of ions was performed in negative mode. High resolution ESI mass spectrometry was performed on a Bruker Apex IV, FT-ICR 7 Tesla mass spectrometer.

II.4.4. NMR analysis

For NMR analysis novclobiocin 401 was dissolved in CD_3OD and NMR spectra were recorded on Varian Inova spectrometers. ^1H NMR spectra were recorded at 600 MHz and ^{13}C NMR spectra were recorded at 125.7 MHz respectively. Chemical shifts in CD_3OD are reported as δ values relative to respective solvent as an internal reference.

III. RESULTS

III.1. Generation and activity test of novclobiocin 401, a clorobiocin derivative containing the catechol moiety 3,4-dihydroxybenzoic acid

III.1.1. Investigation of the substrate tolerance of different aminocoumarin acyl ligases for acyl substrates with catechol moieties

The main aim of this study was the replacement of the genuine 3-dimethylallyl-4-hydroxybenzoyl moiety (Ring A) of clorobiocin (**Figure I.1**) with an acyl moiety containing a catechol structure. We considered six different acids which contain catechol motifs (**Figure III.1A**). Previous mutasynthetic experiments aimed at the replacement of Ring A by other acyl groups, had shown that the success depends primarily on the acceptance of the acyl substrate by the aminocoumarin acyl ligase (or amide synthetase) enzyme that attaches the acyl moiety to the 3-amino group of the aminocoumarin moiety[43, 4]. The later biosynthetic steps, glycosylation and final tailoring steps, appeared not to be affected by modifications of the structure of the acyl moiety. We therefore tested the acceptance of the six acyl substrates with catechol motifs (**Figure III.1A**) by four different aminocoumarin acyl ligases from the biosynthetic gene clusters of novobiocin, clorobiocin, coumermycin A_1, and simocyclinone D8 (NovL, CloL, CouL, and SimL respectively). These four enzymes were expressed and purified as described previously[43, 69, 103, 110], and used for *in vitro* assays of aminocoumarin acyl ligase activity using 3-amino-4,7-dihydroxy-8-methyl-coumarin as amino substrate and the six catechols as acyl substrates. The formation of the amide bond was analysed by HPLC and the identity of the resulting compounds was confirmed by LC-MS. As shown in **Figure III.1**, the best accepted catechol substrate was 3,4-dihydroxybenzoic acid (3,4-DHBA), and the most efficient amide synthetase was CloL, which reached 42% of the reaction velocity observed with the genuine substrate Ring A (**Figure III.1B**). Caffeic acid was not accepted by CloL, but by SimL and NovL albeit with very low efficiency. The other four catechols were not accepted by any of the investigated enzymes. We therefore concentrated our efforts on 3,4-DHBA and the aminocoumarin acyl ligase of clorobiocin biosynthesis, i.e. CloL.

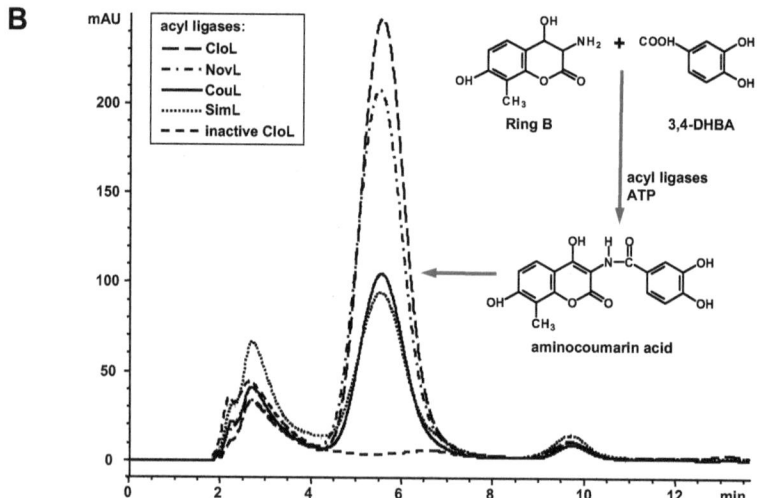

Figure III.1:

(A) Activity of different aminocoumarin acyl ligases (amide synthetases) with acyl substrates containing catechol motifs. Amide bond formation was determined with 3-amino-4,7-dihydroxy-8-methyl-coumarin as amino substrate in the presence of ATP and Mg^{2+}, as described in Materials and Methods. **(B)** HPLC chromatogram showing the product formation of the aminocoumarin acid with 3,4-DHBA under the catalysis of different aminocoumarin acyl ligases.

III.1.2. Inactivation of *cloQ* in the biosynthetic gene cluster of clorobiocin, and heterologous expression of the modified gene cluster

In order to replace Ring A in clorobiocin with 3,4-DHBA, the biosynthesis of Ring A had to be abolished to avoid competition between the genuine and the artificial acyl moiety as precursor in the antibiotic biosynthesis. The first step in Ring A biosynthesis is the prenylation of 4-hydroxyphenylpyruvate catalysed by CloQ[93]. Therefore, we chose to inactivate *cloQ* in cosmid clo-BG1[31], which contains the entire biosynthetic gene cluster of clorobiocin and elements for stable integration in the *Streptomyces* chromosome by RED/ET-mediated recombination[47] to generate cosmid clo-SA2 (see II.2.8.). This *cloQ* deficient gene cluster was integrated into the genome of S. *coelicolor* M512 at the ΦC31 attachment site, resulting in S. *coelicolor* M512(clo-SA2).

Cultivation of this strain in clorobiocin production medium did not result in production of clorobiocin, while similar experiment with S. *coelicolor* M512 containing the intact gene cluster clo-BG1 resulted in the expected production of clorobiocin[31, 37]. Feeding of Ring A restored production of clorobiocin, as confirmed by HPLC, LC-MS analysis and comparison with an authentic reference compound. This demonstrated that the inactivation of *cloQ* led to the abolishment of the production of Ring A, but had not affected the subsequent steps of clorobiocin biosynthesis (**Figure III.2**).

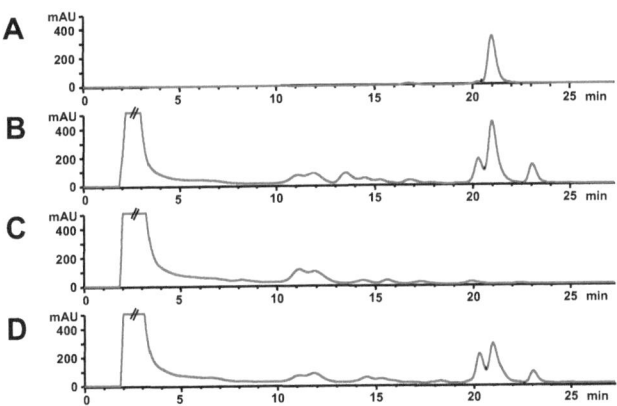

Figure III.2:
HPLC analysis of the *cloQ* defective S. *coelicolor* M512(clo-SA2). **(A)** 1 mM clorobiocin standard; **(B)** analysis of clorobiocin production in S. *coelicolor* M512(clo-BG1); **(C)** abolishment of clorobiocin production in the *cloQ* defective strain S. *coelicolor* M512(clo-SA2); **(D)** restored production of clorobiocin after complementation of S. *coelicolor* M512(clo-SA2) by feeding of 1 mg Ring A.

III.1.3. Mutasynthetic experiments with 3,4-DHBA and caffeic acid

Feeding of different amounts of 3,4-DHBA to cultures of *S. coelicolor* M512(clo-SA2) did not lead to the formation of a clorobiocin derivative, as shown in HPLC and LC-MS analysis. In parallel experiments, we also fed caffeic acid to *S. coelicolor* M512 expressing clo-SA2 and the SimL expression plasmid pSH2[4], since SimL but not CloL was able to accept caffeic acid (**Figure III.1A**). These experiments were also unsuccessful, but we noticed that caffeic acid was clearly detected in the cultures after feeding, while no detectable amounts of 3,4-DHBA were present in the cultures already one day after feeding. We speculated that 3,4-DHBA may rapidly be oxidised in the medium, or quickly metabolised by catabolic pathways similar to those described in *Corynebacterium*[78]. We therefore considered whether a continuous *in vivo* biosynthesis of 3,4-DHBA might be more efficient than external feeding of this compound, in order to supply 3,4-DHBA for aminocoumarin antibiotic formation.

III.1.4. Creating an artificial pathway to 3,4-dihydroxybenzoic acid

3,4-DHBA can be formed by the well characterised 4-hydroxybenzoate-3-hydroxylase PobA of *Corynebacterium cyclohexanicum* (**Figure III.3**)[41, 57]. This 44 kDa flavoprotein monooxygenase is involved in catabolic processes, e.g. lignin degradation. In all Gram-positive bacteria, it uses NADH to provide the required reduction equivalents. The substrate of PobA, 4-hydroxybenzoic acid (4-HBA), is not expected to be present in *Streptomyces* in high concentrations. 4-HBA is an intermediate of ubiquinone biosynthesis. In most organisms, it is formed by degradation of tyrosine[77]. However, Gram-negative bacteria such as *E. coli* synthesise 4-HBA directly from chorismate by elimination of the enol-pyruvyl side chain under catalysis of chorismate pyruvate lyase, a 19 kDa protein encoded by the gene *ubiC* (**Figure III.3**)[106]. The reaction does not require cofactors. By heterologous expression of *ubiC* in the chloroplasts of plants, which do not contain an ortholog of this gene, very large amounts of 4-HBA could be generated without any detrimental effect on growth[118]. Also *Streptomyces* genomes do not contain an ortholog of *ubiC*. An expression of both *ubiC* and *pobA* therefore presented an attractive possibility to generate 3,4-DHBA in our *Streptomyces* strains *in vivo*.

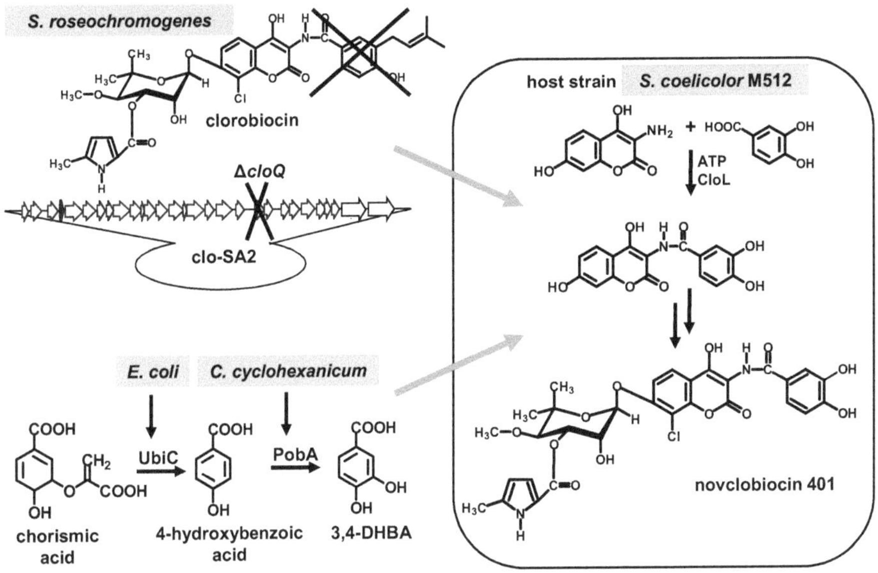

Figure III.3:

Strategy for the generation of a clorobiocin derivative containing a 3,4-dihydroxybenzoyl moiety. Cosmid clo-SA2 contains the entire clorobiocin gene cluster with *cloQ* inactivated.

The GC content of *ubiC* of *E. coli* is 53%, much lower than 70%, the approximate average content in genes of the GC-rich *Streptomyces*. Therefore, the sequence of *ubiC* was modified in order to adapt it to the codon preference of *Streptomyces*. The Gene Designer program (DNA 2.0, USA) was used for this purpose, using the codon usage table of *Streptomyces coelicolor*. The same was applied for the gene *pobA* of *Corynebacterium cyclohexanicum*, although fewer modifications were required for this gene (*Corynebacterium* is also a GC-rich species). The two genes were translationally coupled in order to increase translation efficiency and to facilitate co-regulation. The nucleotide sequence of the synthetic *ubiC/pobA* construct is given in **Figure III.4**. After *in vitro* synthesis (DNA 2.0, USA), this DNA fragment was cloned into the *Streptomyces* expression vector pUWL201[28], placing it under control of the strong constitutive *ermE** promoter. The resulting plasmid, pSA11, was introduced into *S. coelicolor* M512(clo-SA2) by protoplast transformation. Thereby, all genes required for the biosynthesis of the desired compound were assembled in the heterologous expression strain (**Figure III.3**).

```
AAGCTTATGAGCCACCCCGCCCTCACCCAGCTGCGGGCGCTCCGGTACTTCACCGAGATCCCGGCGCTGGAGCCCCAGCT
CCTCGACTGGCTGCTCCTGGAGGACTCGATGACCAAGCGCTTCGAGCAGCAGGGCAAGACCGTCAGCGTCACCATGATCC
GGGAGGGCTTCGTGGAGCAGAACGAGATCCCCGAAGAACTGCCGCTCCTGCCCAAGGAGTCCCGGTACTGGCTCCGGGAG
ATCCTGCTCTGCGCCGACGGGGAGCCCTGGCTGGCCGGCCGGACCGTCGTGCCGGTGTCCACCCTCAGCGGCCCGGAGCT
GGCGCTCCAGAAGCTGGGCAAGACCCCGCTGGGCCGGTACCTCTTCACCTCGTCCACGCTCACCCGGGACTTCATCGAGA
TCGGCCGCGACGCCGGCCTGTGGGGGCGCCGCAGCCGCCTGCGCCTCAGCGGGAAGCCCCTGCTGCTGACGGAGCTGTTC
CTGCCGGCGTCCCCGCTGTACTGATGGGGGACCGGACCGTCATCACCACGCAGGTCGCGATCATGGGTGCGGGTCCGGCC
GGGCTCATGCTGTCCCACCTCCTCCACCAGGCCGGTATCGAAAACACGGTGGTGGAGATCCGGTCCCGCGCGGAGATCTC
CGCCACCATCCGGGCCGGCATCCTGGAGGCCGGTTCGGTCGACCTGCTGGTCCAGAGCGGCGTCGACAACGTCCTCCGGA
ACGGCCACGAACACGAAGGCACCGAGTTCCGCGTCAACGGCGAGGGCCACCGCATCGACTTCAAGGGCCTGGTCGGGCAG
AGCGTCTGGCTGTACCCCCAGAACGACGTCTTCGACGACCTCGCCGCCCGGCGCGAAACGGACGGCGGCGACGTCCGCTA
CTCGTGCAGCAACACCGAGGCCTTCGACCTGCTCGACAAGCCCCGCGTCCACTTCACCGACAGCGAAGGGAACGACTTCG
AACTCCGCGCCGAGATCCTGGTCGGGGCCGACGGGTCCCGGAGCTACTGCCGGCACCAGATCCCCGAAGCCGCCCGGAAG
ACCTACTTCAACGAGTACCCGTTCGCCTGGTTCGGCATCCTGACCGAGGCGCCGCGCAGCGCGCCCGAGCTCATCTACGC
CAACAGCCCCCACGGCTTCGCCCTGATCAGCCAGCGCACCGACACCGTCCAGCGGATGTACTTCCAGTGCGACCCCACCA
CCAACCCCGCCGACTGGACCGACGAGCAGATCTGGGAGCAGCTCCGCCTGCGCGTCAACGGTAACGGTTTCGAGCTCAAG
GAGGGCCCCGTGACCGACAAGGTCGTGCTGCCCTTCCGGTCCTTCGTGCAGACCCCGATGCGCCACGGCAACCTCTTCCT
CGCCGGCGACGCCGCGCACACGGTGCCGCCGACCGGGGCGAAGGGCCTCAACCTGGCCTTCTCGGACGTCCGCGTGCTGT
TCGAATCGCTGGACAGCTACTTCAAGAGCGGCTCCACGGCCCTCATGGACACCTACTCCGAGCGCGCGCTGGACCGGGTG
TGGAAGGCGCAGTACTTCTCGTACTGGATGACCACCCTGCTGCACACCGTGCCGACCGAGACGAACCACGAATTCTTCCG
CGCCCGGCAGCTGGGGGAGCTCCGCTCGCTGCTCGAGTCGGAGCGCGGTCGGGCCTACATCGCCGAGTGCTACACCGGCT
GGCAGTCCAAGTGAACTAGT
```

Figure III.4:

Nucleotide sequence of the synthetic *ubiC/pobA* DNA fragment. Underlined letters represent the flanking *Hind*III and *Spe*I restriction sites. The bold letters show the start and stop codons of the synthetic genes *ubiC* and *pobA*. The genes are translationally coupled, with the stop codon of *ubiC* overlapping the start codon of *pobA*.

III.1.5. Production of novclobiocin 401 by *S. coelicolor* M512(clo-SA2) harbouring plasmid pSA11

S. coelicolor(clo-SA2) harbouring pSA11 was cultivated in three different media: i) the complex clorobiocin production medium described for the wild type producer strain *Streptomyces roseochromogenes* var. *oscitans* DS 12.976[72]; ii) the complex GYM medium described for *S. coelicolor* A(3)2 fermentation[105]; iii) the chemically defined medium (CDM) developed for novobiocin production in the wild type producer strain *Streptomyces niveus*[60]. Ethyl acetate extracts from cultures in three media were analyzed by HPLC and LC-MS. The formation of a new compound with the expected molecular ion of *m/z* 643 [M-H]⁻ was detected in all media. The compound was not detected in extracts from *S. coelicolor*(clo-SA2) lacking plasmid pSA11. The amount of this compound was moderate using the two complex media (4 µg/ml and 7 µg/ml, respectively), but was much higher using the chemically defined medium (64 µg/ml). In this case, the new metabolite clearly represented a major compound in the extract (**Figure III.5A**).

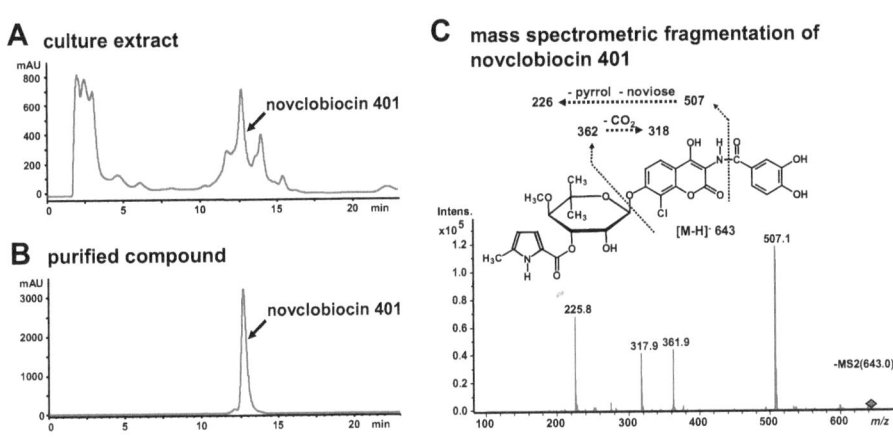

Figure III.5:
(A) HPLC chromatogram of a culture extract from the heterologous producer strain *Streptomyces coelicolor* M512(clo-SA2)/pSA11 cultivated in CDM medium. **(B)** Purified compound isolated from cultures in CDM medium. **(C)** Mass spectrometric fragmentation (MS-MS) of novclobiocin 401 obtained by selected ion monitoring chromatograms. LC-ESI-MS mass scans were performed in negative mode for m/z 643. The suggested fragmentation scheme for the compound is shown.

Preparative isolation of this metabolite from 2 l of culture resulted in 18 mg of pure compound (**Figure III.5B**). This new clorobiocin derivative was termed novclobiocin 401.

III.1.6. Structure elucidation of novclobiocin 401

LC-MS analysis in negative mode showed the presence of the molecular ion of m/z 643 [M-H]⁻, corresponding to the expected molecular mass of 644 for novclobiocin 401. MS/MS analysis (**Figure III.5C**) showed a fragmentation pattern corresponding to that identified in previous mass spectrometric studies of aminocoumarin antibiotics[58]. High resolution ESI-MS in positive mode showed a molecular ion of m/z = 645.14858, which is in agreement with the calculated value of m/z = 645.14818 ($C_{30}H_{30}N_2O_{12}Cl$ [M+H]⁺; Δ0.62 ppm).

Unidimensional (¹H NMR, ¹³C NMR) and multidimensional (¹H ¹H COSY, ¹H HSQC and ¹H HMBC) NMR spectroscopy confirmed the expected structure (**Figure III.6**). Chemical shifts for the substituted pyrrole, the deoxysugar and the aminocoumarin moiety were in accordance to those of clorobiocin[32].

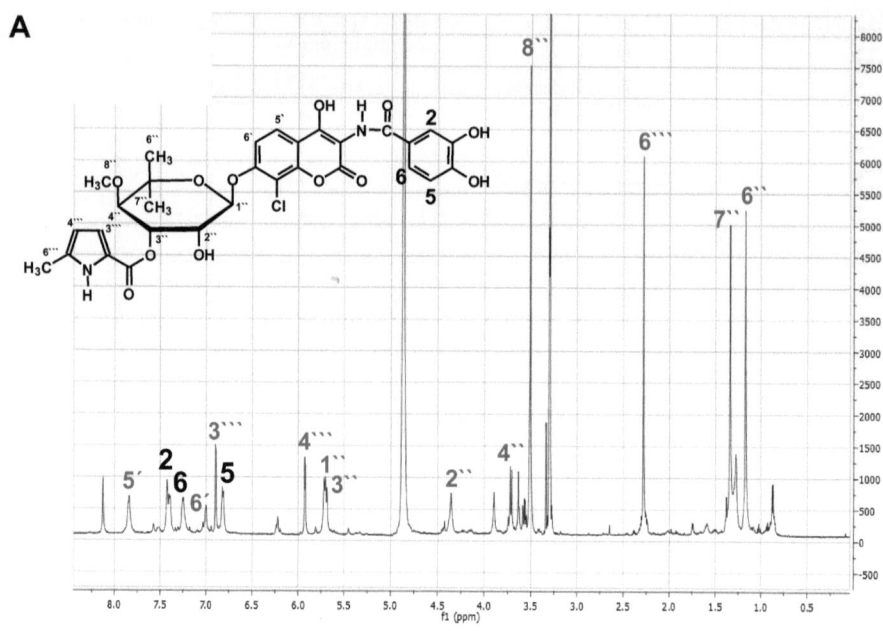
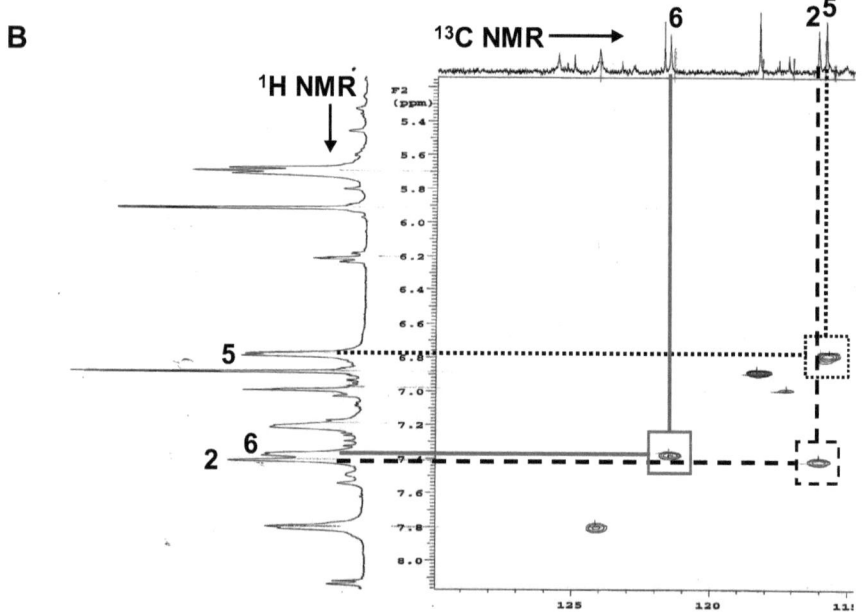

Figure III.6:
^{1}H NMR **(A)** and ^{1}H HSQC **(B)** for the structure confirmation of novclobiocin 401.

RESULTS

The chemical shifts and coupling patterns of the protons in ^{1}H NMR spectra of novclobiocin 401 showed the catechol structure of the new acyl moiety. A strong high-field ^{13}C NMR shift for carbon C-2 (δ_C = 116.1) and C-4 (δ_C = 150.8) in novclobiocin 401 compared to clorobiocin (δ_{C-2} = 130.9, δ_{C-4} = 161.0) indicated a new hydroxyl substitution in ortho-position (C-3). Additionally, the NMR spectra of novclobiocin 401 lack all proton and carbon NMR signals of the dimethylallyl group of the native clorobiocin. Full comparative ^{1}H and ^{13}C NMR spectroscopic data of novclobiocin 401 and clorobiocin are given in **Table III.1**.

Position	^{1}H-NMR data (150.8 MHz, CD$_3$OD), δ_H [ppm], intensity, m, J [Hz]		^{13}C-NMR data (125.7 MHz, CD$_3$OD), δ_C [ppm]	
	novclobiocin 401	clorobiocin	novclobiocin 401	clorobiocin
1			125.6	124.2
2	7.42, 1H, br s	7.76, 1H, d (2.5)	116.1	130.9
3			146.2	129.9
4			150.8	161.0
5	6.80, 1H, d (6.9)	6.84, 1H, d (8.4)	115.8	115.6
6	7.39, 1H, d (6.9)	7.72, 1H, dd (8.4, 2.5)	121.7	128.5
7	--	3.34, 2H, d (7.1)	--	29.2
8	--	5.35, 1H, br t (7.1)	--	123.2
9			--	133.8
10	--	1.74, 1H, s	--	17.9
11	--	1.75, 3H, s	--	26.0
7 in 1, 12 in 2			169.6	170.0
2'			161.8	162.4
3'			102.5	103.7

Position	^{1}H novclobiocin	^{1}H clorobiocin	^{13}C novclobiocin	^{13}C clorobiocin
4'			161.7	157.8
5'	7.82, 1H, d (7.1)	7.90, 1H, d (9.2)	124.1	123.9
6'	7.22, 1H, d (7.1)	7.33, 1H, d (9.2)	111.9	112.5
7'			150.0	161.8
8'			110.2	110.7
9'			156.1	156.5
10'			110.5	113.3
1''	5.72, 1H, d (2.0)	5.73, 1H, d (1.8)	100.2	100.4
2''	4.38, 1H, br m	4.34, 1H, dd (2.7)	71.0	71.0
3''	5.70, 1H, dd (11.1, 2.0)	5.71, 1H, dd (10.3, 2.9)	71.6	71.6
4''	3.72, 1H, d (11.1)	3.72, 1H, d (10.3)	82.7	82.7
5''			80.4	80.5
6''	1.17, 3H, s	1.18, 3H, s	22.9	22.9
7''	1.35, 3H, s	1.35. 3H, s	29.3	29.3
8''	3.52, 3H, s	3.52, 3H, s	62.0	62.1
2'''			121.5	121.8
3'''	6.89, 1H, d (3.4)	6.90, 1H, d (3.6)	118.2	118.33
4'''	5.93, 1H, d (3.4)	5.94, 1H, br d (3.6)	109.7	109.8
5'''			136.1	136.3
6'''	2.29, 3H, s	2.29, 3H, s	13.0	12.9
7'''			161.8	161.8

Table III.1:
Full comparative ^{1}H and ^{13}C NMR spectroscopic data of novclobiocin 401 and clorobiocin. Chemical shifts are expressed in δ values using the solvent as internal standard (CD$_3$OD). Assignments for novclobiocin 401 were made using 2D NMR data; assignments of clorobiocin were taken from literature[32].

III.1.7. Inhibitory activities against *E. coli* and *S. aureus* DNA gyrase and topoisomerase IV.

The new compound was investigated *in vitro* for its inhibitory effects on *E. coli* and *S. aureus* DNA gyrase and topoisomerase IV in comparison with the natural antibiotics clorobiocin and novobiocin. Two different assays were used: a DNA gyrase supercoiling assay and a topoisomerase IV decatenation assay. The 50% inhibitory concentration (IC$_{50}$) of the catechol compound was determined as 0.006 µM for *S. aureus* DNA gyrase and 0.03 µM for *E. coli* DNA gyrase (**Figure III.7A**). These inhibitory concentrations are identical to those observed with clorobiocin and lower than those observed with the clinically used novobiocin (inhibitory concentration: 0.01 µM and 0.08 µM respectively) against the two gyrases. This clearly proves that the replacement of the genuine Ring A moiety with a 3,4-DHBA moiety had not affected the potency of the compound as gyrase inhibitor, what is in agreement with the structural information indicating that Ring A is not involved in the interaction with DNA gyrase (see Introduction).

DNA gyrase inhibitors of the fluoroquinolone class also inhibit topoisomerase IV which is very similar to gyrase[74]. Thus, we examined whether novclobiocin 401 also inhibits the decatenation activity of *E. coli* and *S. aureus* topoisomerase IV. However, just as novobiocin and clorobiocin, the new compound inhibited topoisomerase IV from *S. aureus* and *E. coli* only at much higher concentrations than those required for gyrase inhibition (IC_{50} values: 35 µM for *S. aureus* gyrase and >50 µM for *E. coli* gyrase) (**Figure III.7B**). Therefore, novclobiocin 401 is expected to act primarily as gyrase inhibitor in both *S. aureus* and *E. coli*.

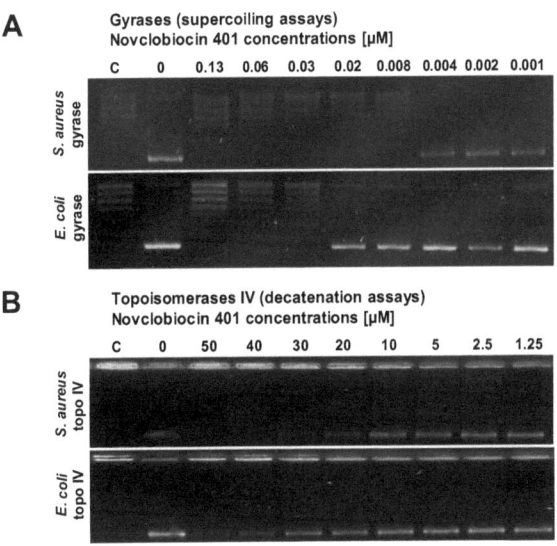

Figure III.7:
Supercoiling **(A)** and decatenation **(B)** assays with *E. coli* and *S. aureus* DNA gyrase and topoisomerase IV. The first lane labelled with C contains control assays without enzyme. The lane labelled 0 contains assays with addition of 3 µl solvent (5% aqueous DMSO). The following lanes contain assays to which the indicated amount of antibiotic, dissolved in 5% DMSO, has been added. In the supercoiling assays, the lower band shows supercoiled pBR322 DNA, formed under catalysis of DNA gyrase. In the decatenation assays, the lower band shows decatenated kinetoplast DNA, formed under catalysis of topoisomerase IV.

III.1.8. Construction of *E. coli* mutants for investigation of antibiotic import by catechol siderophore transporters

As reported above, *E. coli* gyrase is equally sensitive to clorobiocin and novclobiocin 401, but novclobiocin 401 has been designed for a more efficient transport inside the cell than clorobiocin and therefore exert higher antibacterial activity *in vivo*. In order to study the differences in influx without interference by differences in efflux, we inactivated the gene *tolC*, which encodes an essential part of the AcrABTolC drug efflux pumps of *E. coli* (**Figure III.8A**)[15].

E. coli possesses three outer membrane transporters for active import of catechol siderophores: Fiu, Cir, and FepA[18]. All three transporters receive the energy for active transport from the periplasmic-located protein TonB. Therefore, the role of the catechol siderophore transporters Fiu, Cir, and FepA in the uptake of novclobiocin 401 can be investigated by comparing the sensitivity to clorobiocin and novclobiocin 401 of mutants with and without active TonB (**Figure III.8A**). Expression of these proteins is up-regulated under conditions of iron starvation. Bacterial growth during an infection in the human body represents a condition of extreme iron starvation[24]. For functional studies of catechol siderophore transport in *E. coli*, conditions of iron starvation are usually created by i) mutation of *entC*, which abolishes the biosynthesis of the important siderophore enterobactin[26]; ii) addition of the iron chelator 2,2'-bipyridyl to the culture medium[2, 79]. In order to investigate the importance of the outer membrane as a resistant mechanism of Gram-negative bacteria, we considered to test the three aminocoumarins also with *E. coli* strains defective in the outer membrane. A mutation in the gene *rfaP*, encoding for a lipopolysaccharide kinase that is essential for the outer membrane stability[127], causes a hyperpermeable membrane for lipophilic antibiotics by mutational alteration of its lipopolysaccharide components (**Figure III.8A**).

Single in-frame knockout mutants of *entC*, *tolC*, *tonB* and *rfaP* in *E. coli* K-12 were available from the Keio collection[8]. Using these strains, we generated several *E. coli* double and triple mutants by RED/ET-mediated recombination, shown in **Figure III.8B**, to compare their sensitivity to clorobiocin and novclobiocin 401 in culture media containing 2,2'-bipyridyl. The correct genotype of the mutants was verified by PCR analysis (Colony PCR) of the genomic DNA of the mutant.

A

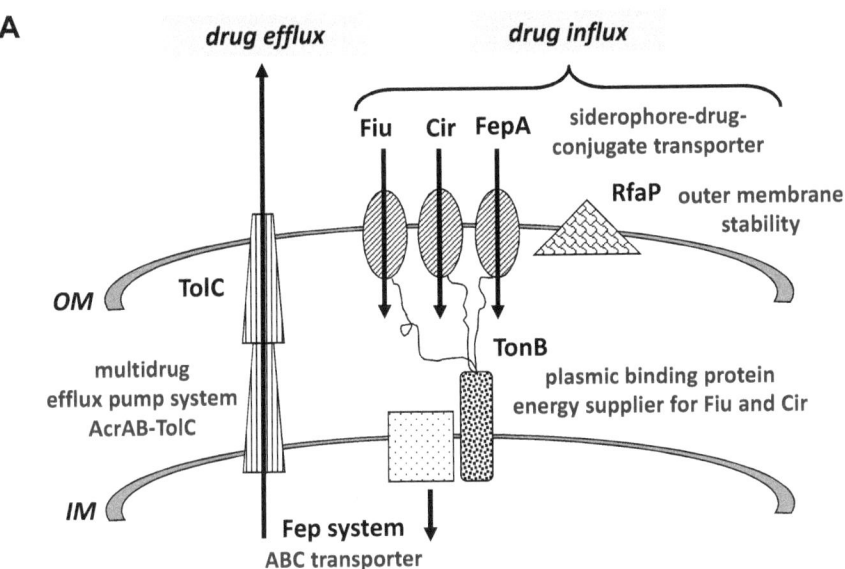

B

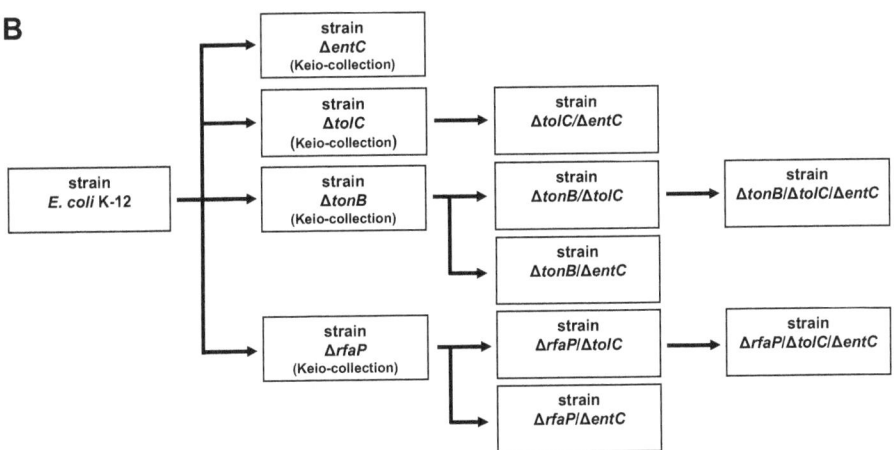

Figure III.8:

(A) Overview of the catechol-mediated drug influx and AcrAB-TolC-mediated drug efflux in *E. coli*. Cir and Fiu are TonB dependent receptors in the outer membrane (OM). In an energy dependent way they catalyse the uptake of siderophore-drug-conjugates. The ABC transporter (Fep system) transports the siderophore-drug-conjugate across the inner membrane (IM) into the cytoplasm. The lipopolysaccharide kinase RfaP is essential for the outer membrane stability; inactivation of *rfaP* results in a mutant which is hypersensitive to aminocoumarins[114, 127]. **(B)** Overview on the constructed multiple mutations of *E. coli* strains by RED/ET-mediated recombination based on strains from the Keio-collection[8].

III.1.9. Determination of the antibacterial activity of novclobiocin 401 in agar diffusion tests

The antibacterial activity of novclobiocin 401, in comparison to clorobiocin and novobiocin, was first tested in agar diffusion assays with the *E. coli* ΔtolC and *E. coli* ΔtolC/ΔentC mutants compared with the *E. coli* ΔtonB/ΔtolC and *E. coli* ΔtonB/ΔtolC/ΔentC mutants. To improve visualisation of the inhibition zones, living cells on the agar plates were stained with 2,3,5-triphenyltetrazolium chloride. The results are shown in **Figure III.9AB**. Against the mutants with intact TonB, and therefore functional siderophore uptake, novclobiocin 401 shows approximately 2-fold higher activity than clorobiocin, and 4-fold higher activity than novobiocin. In contrast, against the *E. coli* mutants with deleted *tonB*, novclobiocin 401 is approximately 10-fold less active than clorobiocin. As expected the antibacterial activity is more pronounced with an additional *entC* mutation, because the production of the own siderophore enterobactin is abolished and therefore bacteria grow under iron starvation and are more sensitive to drugs.

These results show that TonB-dependent transport plays an important role in the antibacterial activity of novclobiocin 401 but not in the activity of clorobiocin and novobiocin, what indicates that i) aminocoumarin antibiotics are efficiently introduced inside *E. coli* and the previously observed resistance[114] is due mainly to the multi-drug efflux pump of which TolC is an important component; and ii) the exchange of the natural Ring A of clorobiocin with 3,4-DHBA diminishes the uptake of the antibiotic through the usual path for clorobiocin and novobiocin. This is in accordance with former literature that Ring A plays a crucial role for the uptake into bacterial cells[1, 98].

Agar diffusion assays with an *E. coli* ΔrfaP/ΔentC mutant support the hypothesis that the outer membrane is not the main resistance mechanism of Gram-negative bacteria against aminocoumarins. The antibacterial activity of aminocoumarins is much more increased in bacteria with an efflux pump defect (*E. coli* ΔtolC/ΔentC mutant) (**Figure III.9B**) than with an outer membrane defect (*E. coli* ΔrfaP/ΔentC mutant) compared with an *E. coli* ΔentC mutant which is not affected at all by aminocoumarins (**Figure III.9C**). However, it remains unclear why the ΔrfaP mutant (supposed to have a hyperpermeable membrane) shows higher resistance to novclobiocin 401 than to clorobiocin or novobiocin (indistinct inhibition zone of 8 mm with 10 µg compound; **Figure III.9C**). A possible explanation is given in section III.1.10.

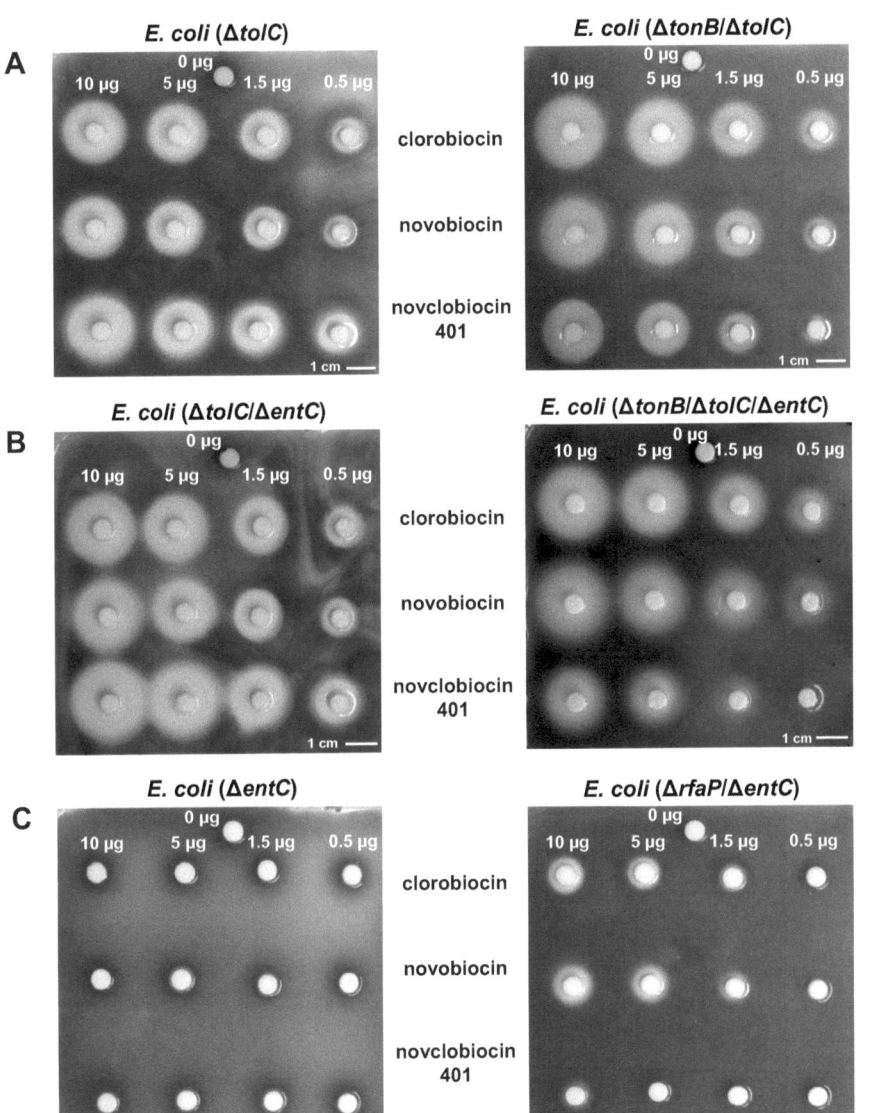

Figure III.9:

Disk diffusion assay for the determination of the antibacterial activity of clorobiocin, novobiocin, and novclobiocin 401 against *E. coli* mutants with and without active TonB **(A B)** and with and without active RfaP **(C)**. At the same time the importance of TolC for the antibacterial activity of aminocoumarins is shown. To improve visualization of the inhibition zones, living cells were stained with 2,3,5-triphenyltetrazolium chloride.

Table III.2 shows the mean values, from at least three separate experiments, of the inhibition zones obtained with different amounts of novobiocin, clorobiocin and novclobiocin 401 (0.5 µg, 1.5 µg, 5 µg, 10 µg) against the described *E. coli* mutants in agar diffusion tests.

E. coli strain	mean of inhibition zone (mm) caused by test compounds											
	0.5 µg			1.5 µg			5 µg			10 µg		
	nov	clo	401	nov	clo	401	nov	clo	401	nov	clo	401
Δ*tolC*	9	9.3	14	11	13.8	16.3	15.5	17.5	19.3	18.3	19.2	21
Δ*tolC*/Δ*entC*	10.5	10.6	15.3	12.5	15.2	21.3	18.3	18.4	25.6	21	19.8	26
Δ*tonB*/Δ*tolC*	10.3	11.3	8	14	15.3	12.6	18.6	19.6	17.3	22.6	21.6	19.6
Δ*tonB*/Δ*tolC*/Δ*entC*	11.8	13	9.5	16	17.5	10	20	21.5	16	22.3	23	19.3
Δ*rfaP*/Δ*entC*	0	0	0	13.3	11.5	0	16.6	13	0	19.6	15.6	8
Δ*entC*	0	0	0	0	0	0	0	0	0	0	0	0
K12 (wild type)	0	0	0	0	0	0	0	0	0	0	0	0

Table III.2:
Inhibition zones obtained with novobiocin (nov), clorobiocin (clo) and novclobiocin 401 (401) against genetically defined *E. coli* mutants in agar diffusion tests, performed in Mueller-Hinton-Agar; tested compounds were dissolved in MeOH; inhibition zones are averages from at least three separate experiments.

After the evaluation of the inhibition zones the statistical t-test was performed to test whether the antibacterial activity of novclobiocin 401 was significantly better than that of clorobiocin or novobiocin. It is generally accepted that if a specific parameter (p value) in the t-test is lower than 0.05 then the antibacterial activities are different with statistical significance, i.e. with 95% probability.

Figure III.10 shows that the antibiotic activities of novobiocin and clorobiocin against the *E. coli* Δ*tolC*/Δ*entC* mutant are not significantly different (p value >0.05). In contrast, novclobiocin 401 showed a significantly better antibacterial activity than novobiocin and clorobiocin against this mutant (p value <0.05). On the other hand, novclobiocin 401 showed a significantly lower activity against the *E. coli* mutants without active TonB than clorobiocin or novobiocin, what demonstrates that Ton B is involved in the uptake of novclobiocin 401 into the Gram-negative cell.

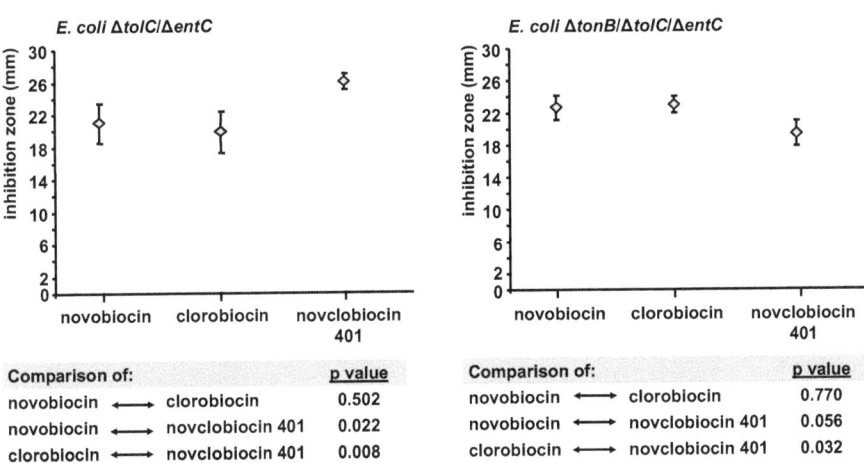

Figure III.10:
Statistical t-test to determine the significant differences in the antibacterial activity of novclobiocin 401 in comparison with novobiocin and clorobiocin. Both graphs represent the mean of inhibition zones caused by the tested compounds (10 µg) against *E. coli* mutants with and without active TonB. Error bars indicate the calculated standard deviations.

III.1.10. Growth promotion with enterobactin

The observation that novclobiocin 401 was not able to enter the hyperpermeable membrane mutant *E. coli* ΔrfaP/ΔentC, unlike clorobiocin or novobiocin, was unexpected. To investigate this further, we performed growth promotion assays with the *E. coli* siderophore enterobactin. We hypothesised that due to the alteration of the lipopolysaccarids (caused by ΔrfaP mutation) the siderophore transporters, located in the outer membrane, might have been also affected. This could be a reason why novclobiocin 401 can not be transported inside the cell and shows less antibacterial activity. Therefore, we compared the *E. coli* ΔentC and *E. coli* ΔrfaP/ΔentC mutants, both unable to form their own siderophore enterobactin, under iron-restricted conditions. After addition of enterobactin to the *E. coli* ΔentC mutant, growth could be observed. In contrast, the *E. coli* ΔrfaP/ΔentC mutant showed less ability to grow on iron diminished medium what points to a lack of iron transport into the cell (**Figure III.11**). This experiment gives a possible explanation of the limited influx of novclobiocin 401 through the hyperpermeable membrane, which like enterobactin is transported by siderophore transporters.

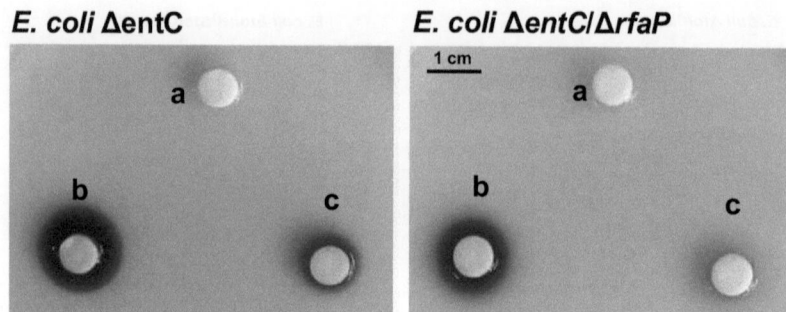

Figure III.11:
Role of RfaP in enterobactin-mediated growth promotion. Both strains were grown in iron-restricted medium. **(a)** control (methanol); **(b)** 0.02 µg enterobactin; **(c)** 0.01 µg enterobactin. Dark halos indicate bacterial growth.

III.1.11. Determination of the minimum inhibitory concentration (MIC) of novclobiocin 401

We also determined the minimum inhibitory concentration (MIC) in liquid culture, following the procedure described by Wiegand and co-workers[123], but including 50 µM of the iron chelating agent 2,2'-bipyridyl to the culture media. Under these conditions, novobiocin and clorobiocin MIC against *E. coli* ΔtolC/ΔentC double mutant was 23 and 12 µg/ml respectively. The MIC of novclobiocin 401 was 6 µg/ml, showing again that novclobiocin 401 has higher antibacterial activity than its parent compound clorobiocin. Against the ΔtonB/ΔtolC/ΔentC triple mutant, novobiocin and clorobiocin gave the same MIC values as observed against the ΔtolC/ΔentC double mutant (23 and 12 µg/ml, respectively). However, novclobiocin 401 showed much less activity against the *tonB*-deficient triple mutant (MIC: 47 µg/ml) than against the double mutant with active *tonB* (MIC: 6 µg/ml). This shows again that TonB-dependent transport is important for the antibacterial activity of novclobiocin 401, but not of novclobiocin and clorobiocin.

Against the wild-type strain *E. coli* K-12, the MICs of novobiocin, clorobiocin and novclobiocin 401 resulted as 375, 47 and 95 µg/ml respectively. Similar values were obtained against the ΔentC mutant. Therefore, against these strains novclobiocin 401 was less active than its parent compound clorobiocin. The poor activity of aminocoumarins against wild-type strains of *E. coli* is in accordance with previous observations[5, 44].

III.2. Generation of a clorobiocin derivative containing the catechol moiety 2,3-dihydroxybenzoic acid

III.2.1. Activation of 2,3-dihydroxybenzoic acid by the AMP ligase DhbE from *Bacillus subtilis*

After the generation of a clorobiocin derivative with the siderophore-like structure 3,4-dihydroxybenzoic acid (novclobiocin 401), the production of a second clorobiocin derivative with 2,3-dihydroxybenzoic acid (2,3-DHBA) was of great interest as the siderophores enterobactin from *E. coli* and bacillibactin from *B. subtilis* contain this catechol motif.

In amide synthetase assays (see III.1.1.) we found that none of the four available acyl ligases (NovL, CloL, CouL, SimL) accepted 2,3-DHBA as substrate. Normally, the acyl ligases catalyze two steps, the adenylation of the acyl moiety and the following attachment to the aminocoumarin moiety.

The required genes for the biosynthesis of 2,3-DHBA, precursors of the siderophore bacillibactin[75], are organized in a single operon in *B. subtilis*. Conspicuous is that a special enzyme, an AMP ligase, is present to catalyze the activation of 2,3-DHBA to 2,3-DHB adenylate (**Figure III.12**).

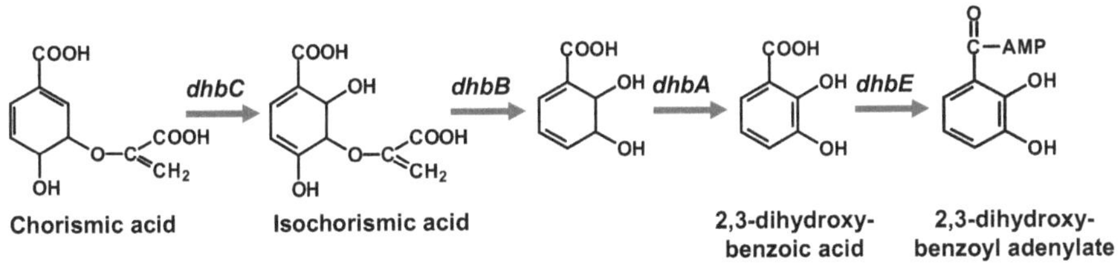

dhbC	Isochorismate synthase
dhbB	Isochorismate lyase
dhbA	Dehydrogenase
dhbE	AMP ligase

Figure III.12:
Enzymes of bacillibactin biosynthesis.

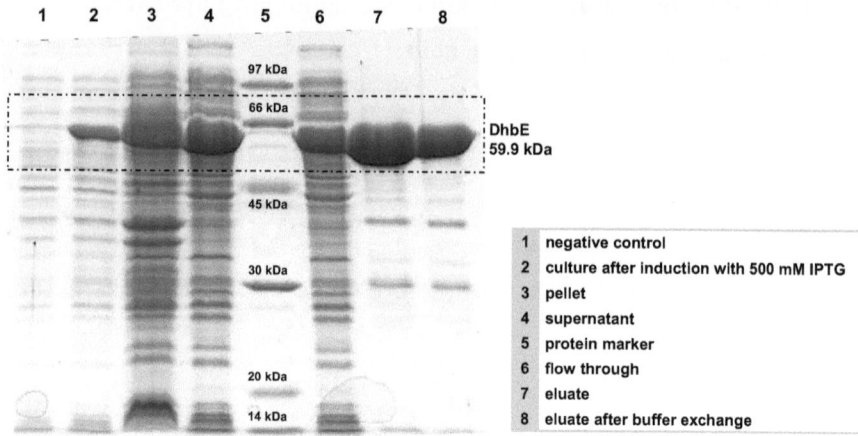

Figure III.13:
SDS PAGE of the expression and purification of the AMP ligase DhbE from *B. subtilis*.

We concluded that 2,3-DHBA needs a special activation, which can not be catalysed by any of the acyl ligases alone. We design the *in vitro* adenylation of 2,3-DHBA to study its acceptance by the for amide synthetases. Plasmid pJJM301, that contains *dhbE* from *B. subtilis*, was kindly provided by the laboratory of Prof. M. Marahiel (Marburg, Germany). Soluble, N-terminally his-tagged protein DhbE was obtained after expression of plasmid pJJM301 in *E. coli* BL21/pREP4 cells. Ni^{2+}-affinity chromatography resulted in approximately 21 mg of purified DhbE from one litre LB-culture. The apparent molecular mass in SDS-PAGE analysis was consistent with the predicted molecular weight of the polypeptide encoded by *dhbE* (59.9 kDa) (**Figure III.13**). We then performed *in vitro* amide synthetase assays as described in section II.3.6 but containing equal amounts of purified DhbE and of one of the acyl ligases per assay. 2,3-DHBA was efficiently adenylated by purified DhbE, and the activated catechol was readily accepted as substrate by the four available amide synthetases and connected to the coumarin moiety (**Figure III.14**).
Therefore, the reason for the failed reaction described in section III.1.1 was the required pre-activation by DhbE (**Figure III.15**).

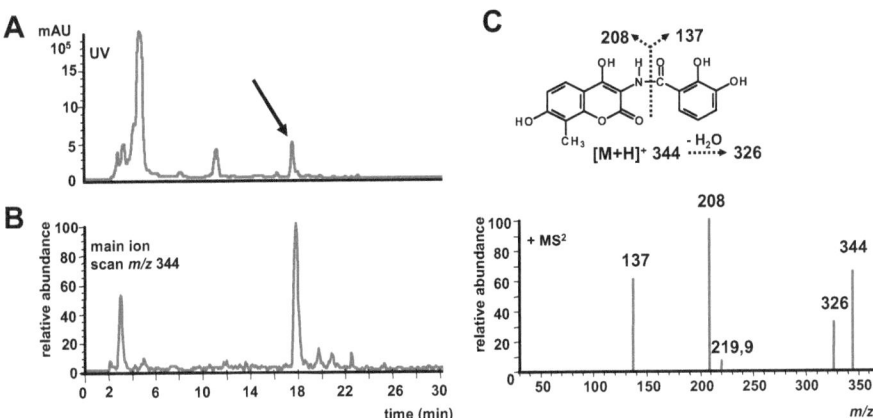

Figure III.14:

(A) HPLC chromatogram of an amide synthetase assay containing CloL and DhbE in equal amounts; the arrow indicates the product formation of the aminocoumarin acid with 2,3-DHBA moiety. **(B)** LC-MS chromatogram showing the MS-MS main ion scan for m/z 344 in positive mode. **(C)** Mass spectrometric fragmentation of the novclobiocin acid with 2,3-DHBA moiety obtained by selected ion monitoring chromatograms. The suggested scheme for the compound is shown.

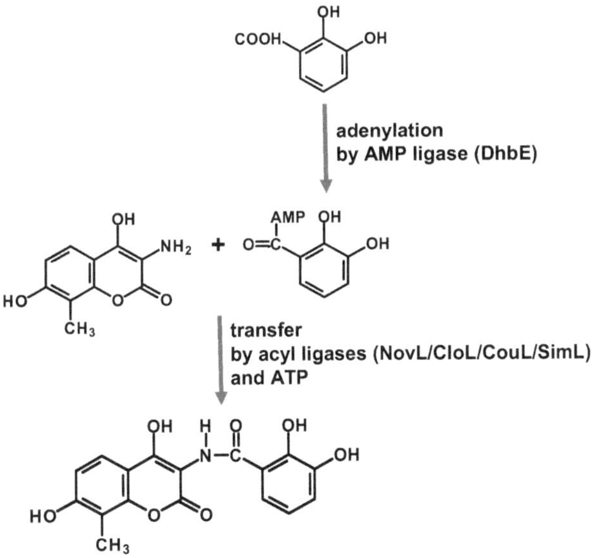

Figure III.15:

Suggested activation of 2,3-DHBA by the AMP ligase DhbE by an adenylation reaction and the following transfer to the aminocoumarin moiety catalyzed by amide synthetases under catalysis of ATP.

III.2.2. Detection of the clorobiocin derivative with 2,3-dihydroxybenzoic acid

S. coelicolor genome contains a gene cluster for a hypothesised zinc-chelating compound, called coelibactin[11, 51]. The first gene in this hypotetical cluster, sco7681, encodes a putative AMP-binding ligase with 52% amino acids homology to DhbE from B. subtilis. We considered that this DhbE homolog could activate 2,3-DHBA in a similar way than DhbE of the catechol siderophore griseobactin found in S. griseus[89].

The first attempts to obtain the new clorobiocin derivative by feeding of 2,3-DHBA to S. coelicolor M512(clo-SA2) remained unsuccessful, although 2,3-DHBA, in contrast to 3,4-DHBA, could be detected over several days in the culture and therefore the problem of rapid degradation of the catechol by Streptomcyes was not the problem. This could be due to lack of expression of sco7681 under our working conditions.

We then tried to synthesise 2,3-DHBA in S. coelicolor M512 by introducing the plasmid pSP126$_{1110}$ containing the necessary genes (dhbA, dhbC, dhbE, dhbB; **Figure III.12**) from griseobactin gene cluster of Streptomcyes griseus[89]. This strategy was also unsuccessful, probably because of inefficient expression of the genes from that construct.

In a further experiment we integrated the cloQ deficient clorobiocin biosynthetic gene cluster (clo-SA2) as well as the plasmid pSP126$_{1110}$ (containing genes for griseobactin biosynthesis) into the genome of S. coelicolor M1154 at the ΦC31 attachment site, resulting in S. coelicolor M1154(clo-SA2)/pSP126$_{1110}$. We cultivated this strain in chemically defined medium (CDM) and additionally fed 2,3-DHBA to the culture. Ethyl acetate extracts from the culture were analyzed by HPLC and LC-MS after 5 d. LC-MS analysis in negative mode showed the presence of the molecular ion of m/z 643 [M-H]$^-$, corresponding to the expected molecular mass of 644 for the new clorobiocin derivative with 2,3-DHBA. MS/MS analysis (**Figure III.16**) showed the expected fragmentation pattern of the new compound. We think that the activation of the externally added 2,3-DHBA was actually performed by the AMP-ligase encoded by sco7681 and not by the DhbE encoded in pSP126$_{1110}$, since the formation of this compound was only detected in the culture to which 2,3-DHBA was added and therefore the genes in pSP126$_{1110}$ seem not to be expressed. S. coelicolor M1154 is a new host strain especially developed for the increased heterologous expression of secondary metabolite gene clusters[45] what could explain the expression of sco7681 in contrast to M512.

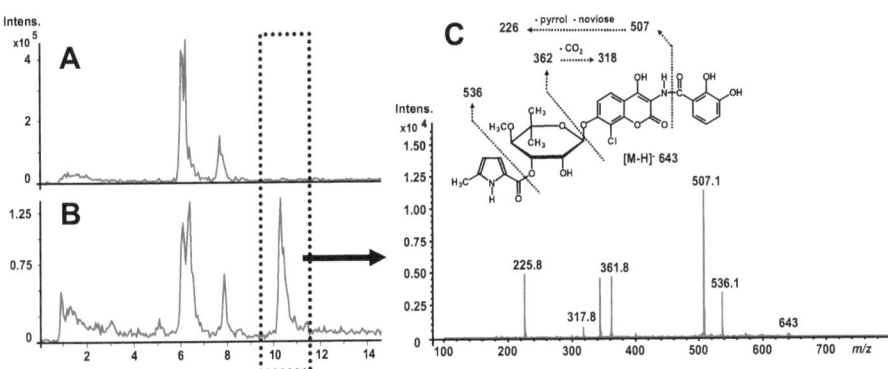

Figure III.16:
LC-ESI-MS mass scans were performed in negative mode for m/z 643. Extracted ion chromatogram for the main ion m/z 643 [M-H]⁻ of the culture extract *S. coelicolor* M1154(clo-SA2)/pSP126$_{1110}$ without feeding of 2,3-DHBA (**A**) and with feeding of 2,3-DHBA (**B**). (**C**) Mass spectrometric fragmentation of the new clorobiocin derivative with 2,3-DHBA obtained by selected ion monitoring chromatograms. The suggested fragmentation scheme for the compound is shown.

Despite we have certainly obtained the desired compound, it is still necessary to optimize the conditions for an enhanced production of the new clorobiocin derivative with 2,3-DHBA, in order to obtain enough amounts for structure confirmation by NMR and bioactivity tests.

III.3. Inhibition of DNA gyrase and topoisomerase IV of *Staphylococcus aureus* and *Escherichia coli* by aminocoumarin antibiotics

III.3.1. Expression of the subunits of *Staphylococcus aureus* topoisomerase IV as his-tagged proteins

S. aureus DNA gyrase, as well as *E. coli* topoisomerase IV and DNA gyrase, are commercially available, but it was necessary to generate *S. aureus* topoisomerase IV for our study. Expression of the structural genes for ParC and ParE as C-terminally his-tagged proteins, using vector pET22b (Novagen) were unsuccessful. However, soluble, N-terminally his-tagged ParC and ParE proteins were obtained from constructs containing *parC* and *parE* from *S. aureus* RN4220 in the pQE70 vector (QIAGEN) which were expressed in *E. coli* BL21/pREP4 cells. Ni^{2+}-affinity chromatography resulted in approximately 7 mg of purified ParC and 5 mg purified ParE per litre of culture. The apparent molecular mass in SDS-PAGE analysis was consistent with the predicted molecular weight of the polypeptides encoded by *parC* (96.2 kDa) and *parE* (77.8 kDa) respectively. Mixing of equimolar amounts of the two subunits resulted in active topoisomerase IV (see below).

III.3.2. Removal of potassium glutamate from the assays for DNA gyrase activity

We established assay conditions for the measurement of the inhibition of *E. coli* and *S. aureus* DNA gyrase and topoisomerase IV by aminocoumarin antibiotics. As described by Morgan-Linnell and co-workers[80] and Pan & Fisher[88] for investigation of the inhibitory activity of agents against topoisomerase IV the decatenation assay is the most appropriate, while the most relevant assay for agents like aminocoumarins, acting on DNA gyrase, is inhibition of supercoiling. The activity of *S. aureus* DNA gyrase is dependent on high concentrations of potassium glutamate (K-Glu)[16, 52]. However, we found that K-Glu concentrations over 400 mM impaired the resolution in the gel electrophoresis analysis of the supercoiling assays (**Figure III.17, lane b**). We tested different methods for desalting the samples before loading them in agarose gels.

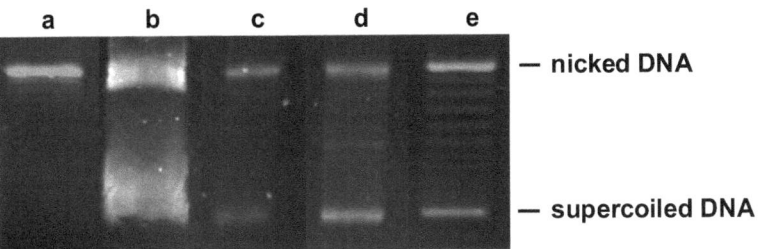

Figure III.17:
Electrophoretic analysis of S. aureus DNA gyrase supercoiling assays using different methods for potassium glutamate removal. Lane **a**, assay without K-Glu (the enzyme is inactive under these conditions); lane **b** to **e**, assays with 700 mM K-Glu; lane **b**, without removal of K-Glu; lane **c**, addition of 700 mM 18-Crown-6; lane **d**, desalting with QIAspin Miniprep Kit for plasmid DNA purification (QIAGEN); lane **e**, dialysis against 10 mM Tris-HCl buffer (pH 8.0) with 0.025 µM membrane filters MF^{TM} VSWP (Millipore).

Addition of 700 mM 18-Crown-6 (Fluka) after the reaction, intended to complex the potassium cations, led to an undesirable reduction of the intensity of the bands on the agarose gel (**Figure III.17, lane c**). Purification of the assay products with the QIAprep Spin Miniprep Kit for plasmid DNA purification (QIAGEN) provided better results (**Figure III.17, lane d**). The best resolution and visualization of the different topoisomers of relaxed DNA was achieved by dialysis against a 10 mM Tris-HCl buffer (pH 8.0) using membrane filters (0.025 µm pore diameter) (**Figure III.17, lane e**), and therefore this method was used for all further DNA gyrase assays. No dialysis was required for topoisomerase IV activity tests. Kinetoplast DNA was used as substrate in these assays, and the resolution in agarose gel electrophoresis was satisfactory without removal of K-Glu (**Figure III.18**). Furthermore, S. aureus topoisomerase IV required a lower concentration of K-Glu for activity than DNA gyrase (see below).

III.3.3. Effect of potassium glutamate on the activity of DNA gyrase and topoisomerase IV of *E. coli* and *S. aureus*

Previous investigations had shown that K-Glu has an important influence on the activities of bacterial type II topoisomerases. The optimal K-Glu concentration is different for *S. aureus* and for *E. coli* enzymes, and different for DNA gyrase and topoisomerase IV. Even for the same enzyme of a given origin, the optimal K-Glu concentration varies for different assay types, e. g. for the supercoiling and the DNA cleavage assay in case of *S. aureus* DNA gyrase, or for the decatenation and the relaxation assay in case of topoisomerase IV[16,52,99]. We investigated the influence of K-Glu and Na-Glu in concentrations from 0 mM to 1000 mM on the four enzymes relevant to our study, i.e. DNA gyrase and topoisomerase IV from both *S. aureus* and *E. coli*, using the supercoiling assay for DNA gyrase and the decatenation assay for topoisomerase IV (**Figure III.18**).

S. aureus DNA gyrase and topoisomerase IV were both inactive in the absence of K-Glu. DNA gyrase activity was detectable from 300 - 1000 mM K-Glu, with an optimal concentration around 900 mM. Topoisomerase IV activity was detectable between 100 - 500 mM K-Glu, with an optimal activity between 200 - 400 mM. These data are in agreement with the results of the four previous studies on the influence of K-Glu on *S. aureus* topoisomerases[16, 52, 99, 112].

E. coli DNA gyrase is active in the absence of K-Glu, and this may be the reason that only few data are available on the influence of K-Glu on type II topoisomerases of *E. coli*[52]. Our investigations confirmed that *E. coli* DNA gyrase does not require K-Glu for activity, but also showed that inclusion of 200 - 500 mM K-Glu moderately stimulates the activity of this enzyme. Using the established amount of enzyme in our standard decatenation assay, activity of *E. coli* topoisomerase IV was not detected in the absence of K-Glu, but clearly in the presence of 100 - 500 mM of this salt; the optimal concentration range was 200 – 400 mM (**Figure III.18**). Therefore, also *E. coli* topoisomerase IV is clearly stimulated by K-Glu. Blanche and co-workers[16] reported that for stimulation of *S. aureus* DNA gyrase in the supercoiling assay, K-Glu could be replaced by its enantiomer, e.g. the potassium salt of D-Glu, but not by Na-Glu or KCl. Our investigations confirmed that *S. aureus* DNA gyrase is inactive in the presence of Na-Glu, irrespective of the concentration (**Figure III.18**). In clear contrast, *S. aureus* topoisomerase IV is stimulated by Na-Glu (200 – 400 mM), although less than by K-Glu. Also *E. coli* topoisomerase IV is weakly stimulated by 200 mM Na-Glu. Notably, *E. coli* DNA gyrase is completely inhibited by Na-Glu in concentrations of 200 mM or higher (**Figure III.18**). It has been shown that a monovalent cation is required for the ATPase activity of this family of enzymes and that it seems to exist some specificity for the cation[56]. To test whether potassium is responsible for the catalytic activity of *S. aureus* DNA gyrase we performed similar series of supercoiling assays with 0 mM to 1500 mM KCl or NaCl.

DNA gyrase (supercoiling assay)
potassium glutamate concentrations:

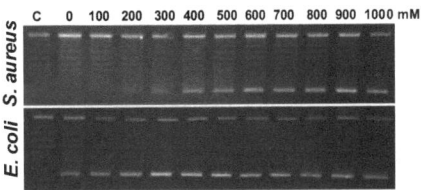

topoisomerase IV (decatenation assay)
potassium glutamate concentrations:

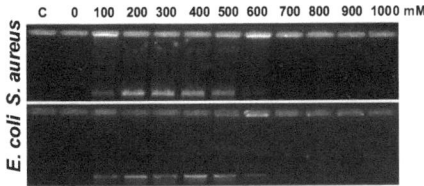

sodium glutamate concentrations:

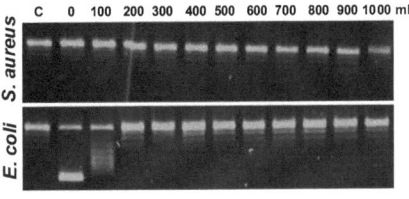

sodium glutamate concentrations:

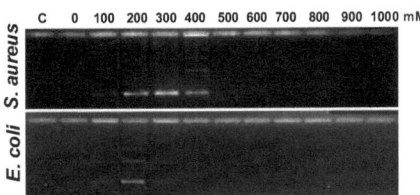

Figure III.18:
Effect of K-Glu and Na-Glu on *S. aureus* and *E. coli* DNA gyrase in supercoiling assays and on topoisomerase IV in decatenation assays. The supercoiling reaction mixtures contained relaxed pBR322 DNA, DNA gyrase and the indicated concentrations of K-Glu and Na-Glu. The decatenation reaction mixtures contained kDNA, topoisomerase IV and the indicated concentrations of K-Glu and Na-Glu. The first lane, labelled with C, contains a control assay without enzyme. In the DNA gyrase assays, the lower band shows supercoiled DNA. In the topoisomerase IV assays, the lower band shows decatenated DNA.

E. coli **DNA gyrase (supercoiling assay)**
KCl concentrations:

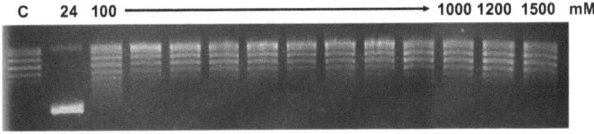

E. coli **DNA gyrase (supercoiling assay)**
NaCl concentrations:

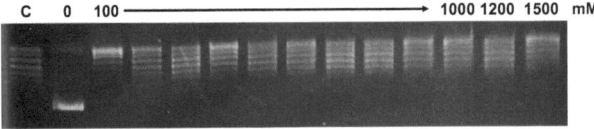

Figure III.19:
Effect of KCl and NaCl on *E. coli* DNA gyrase in supercoiling assays. The first lane, labelled with C, contains a control assay without enzyme. In the DNA gyrase assays, the lower band shows supercoiled DNA.

S. aureus DNA gyrase was inactive regardless the concentration of KCl or NaCl, while the *E. coli* enzyme lost activity from 100 mM onwards of either KCl or NaCl (**Figure III.19**). Since *E. coli* DNA gyrase does not require nor is affected by K-Glu, the lack of activity observed when adding Na-Glu could indicate that the enzyme prefers potassium, and sodium is competing for the binding to the enzyme. To test this hypothesis we performed supercoiling assays in the presence of 100 mM Na-Glu and increasing concentrations of K-Glu up to 400 mM, but K-Glu did not restore enzyme activity (data not shown).

III.3.4. Potassium glutamate modulates the sensitivity of *E. coli* DNA gyrase to aminocoumarin antibiotics

Previous studies have shown that the sensitivity of DNA gyrase to quinolones is modulated by K-Glu[52, 112]. We now tested the influence on K-Glu on the sensitivity of *E. coli* DNA gyrase to aminocoumarin antibiotics. *E. coli* DNA gyrase was approximately 10-fold more sensitive to novobiocin, clorobiocin, and novclobiocin 101 (**Figure III.20**) in the presence of 700 mM K-Glu than in its absence (**Figure III.21**). Unexpectedly, the effect of K-Glu was much more pronounced in case of novclobiocin 103, which is an analogue of novobiocin lacking the acyl substituent, which is attached in the naturally occurring antibiotic (**Figure III.20**) to the 3-OH group of the deoxysugar moiety. *E. coli* DNA gyrase was only weakly inhibited by this compound in the absence of K-Glu, but sensitivity increased by a factor of 150 in the presence of K-Glu. X-ray crystallographic studies have shown that the carbamoyl group of novobiocin occupies a distinct binding pocket of the DNA gyrase protein, providing an important interaction with the target[74]. Apparently, the relative importance of this particular interaction in the overall binding of the antibiotic is reduced in the presence of K-Glu.

The concentration of K-Glu is therefore an important consideration when the inhibitory effects of aminocoumarins on DNA gyrase from *E. coli* and *S. aureus* are compared. Since DNA gyrase from *S. aureus* does not require K-Glu, we decided to subsequently use 700 mM K-Glu in assays of DNA gyrases from both organisms that is the concentration recommended by Blanche and co-workers[16] for assays of *S. aureus* DNA gyrase and still provides 100% of activity of *E. coli* DNA gyrase.

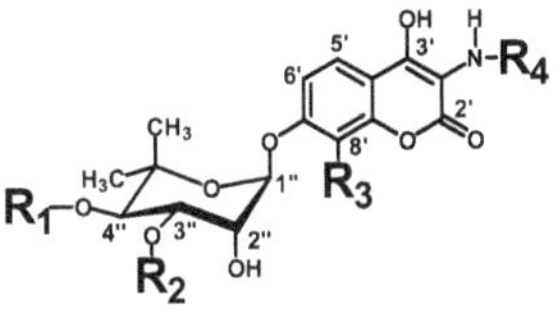

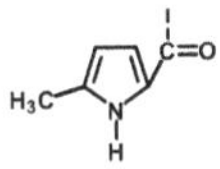

compound	R₁	R₂	R₃	R₄
novobiocin	CH_3	$CONH_2$	CH_3	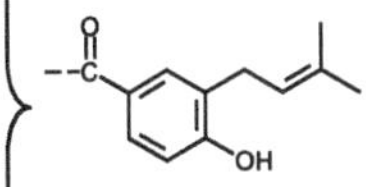
clorobiocin	CH_3	MePC	Cl	
novclobiocin 101	CH_3	MePC	H	
novclobiocin 102	CH_3	MePC	CH_3	
novclobiocin 103	CH_3	H	CH_3	
novclobiocin 105	H	H	Cl	
novclobiocin 217	CH_3	MePC	CH_3	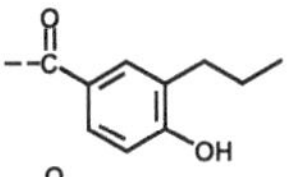
novclobiocin 225	CH_3	MePC	CH_3	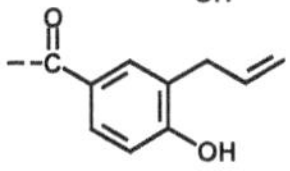

Figure III.20:
Structure of the novclobiocins used in this study.

The other three enzymes in our study, i.e. DNA gyrase from *S. aureus* and topoisomerase IV from both *S. aureus* and *E. coli*, did not show detectable activity in the absence of K-Glu under our assay conditions, and therefore the influence of K-Glu on their sensitivities to aminocoumarins could not be tested in a similar experiment as described above for *E. coli* DNA gyrase. We decided to use 100 mM K-Glu in subsequent assays for topoisomerase IV from both organisms, which is the concentration used in several previous studies[35, 90] and which is recommended by the commercial suppliers of topoisomerase IV, i.e. Inspiralis (Norwich, UK) and Topogen (Columbus, Ohio, USA). However, our own data suggest that the enzyme activity is slightly higher at 200 – 400 mM K-Glu (**Figure III.18**).

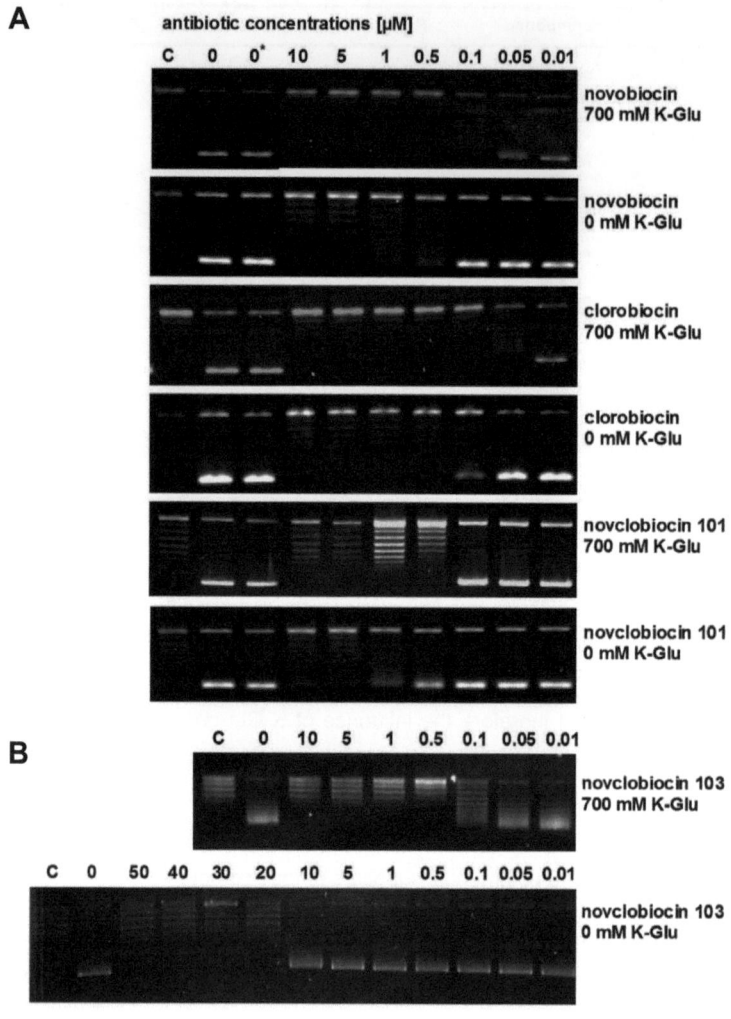

Figure III.21:

Influence of K-Glu on the sensitivity of *E. coli* DNA gyrase towards aminocoumarin antibiotics. The supercoiling reaction mixtures contained relaxed pBR322 DNA, DNA gyrase and the indicated concentrations of the antibiotic. The first lane, labelled with C, contains control assays without enzyme. The lanes labelled 0 or 0* contain assays with no addition or with addition of 3 µl solvent (5% aqueous DMSO), respectively. The following lanes contain assays to which the indicated amount of antibiotic, dissolved in 5% DMSO, has been added. **(A)** Investigation of novobiocin, clorobiocin and novclobiocin 101. **(B)** Investigation of novclobiocin 103.

III.3.5. Inhibition of DNA gyrase and topoisomerase IV from *E. coli* and *S. aureus* by different aminocoumarin antibiotics

Using the assay conditions established in the preceding experiments, we investigated the inhibitory effect of different aminocoumarin antibiotics on DNA gyrase and topoisomerase IV from *E. coli* and *S. aureus*. We tested the three "classical" aminocoumarin antibiotics novobiocin, clorobiocin, and coumermycin A_1, as well as the structurally different simocyclinone D8 (**Figure III.20**). We also included several novobiocin and clorobiocin derivatives (termed "novclobiocins"), which we had obtained in previous mutasynthesis and metabolic engineering experiments[4, 32, 35, 126]. Novobiocin, clorobiocin, and coumermycin A_1 were on average 6-fold more active against *S. aureus* DNA gyrase than against *E. coli* DNA gyrase (**Table III.3**). The inhibitory concentrations were in the range of 6-10 nM, i.e. three orders of magnitude lower than reported for fluoroquinolones like ciprofloxacin, ofloxacin or sparfloxacin[88, 113, 115]. This confirms the potency of aminocoumarins as DNA gyrase inhibitors in Gram-positive pathogens. However, the activity of the aminocoumarins against *S. aureus* topoisomerase IV was weaker (**Table III.3**). These biochemical data suggest that gyrase is the primary target of the tested aminocoumarins in *S. aureus*. This is supported by the observation by Fujimoto-Nakamura and coworkers[42], who showed that cultivation of *S. aureus* in the presence of novobiocin resulted first in the selection of mutants with altered DNA gyrase; use of higher concentrations of novobiocin additionally resulted in mutations of topoisomerase IV as a second step in the emergence of resistance.

Simocyclinone D8, an aminocoumarin antibiotic with a completely new mode of action[30, 36], showed similar activity against *E. coli* DNA gyrase as novobiocin, but less activity against *S. aureus* DNA gyrase. Its effect on *S. aureus* topoisomerase IV was weak and it was essentially inactive against the corresponding *E. coli* enzyme (**Table III.3**)[87].

Novclobiocin 101, which is very similar in structure to clorobiocin but lacks the chlorine in position 8 of the aminocoumarin moiety (**Figure III.20**) showed 8-fold lower inhibitory activity against *S. aureus* DNA gyrase than clorobiocin (**Table III.3**). If the chlorine atom of clorobiocin is replaced by a methyl group (novclobiocin 102), there is 2-fold reduction in activity. These biochemically determined ratios of activities between clorobiocin, novclobiocin 101, and novclobiocin 102 are identical to those determined earlier in a disk diffusion assay against *Bacillus subtilis*[32].

When the acyl substituent at position 3 of the deoxysugar of novobiocin is removed, resulting in novclobiocin 103, activity against *S. aureus* DNA gyrase is reduced by a factor of 100. When the methyl group of the 4-methoxy group at the deoxysugar moiety of clorobiocin is additionally removed, activity is completely lost (novclobiocin 105).

Novclobiocins 217 and 225 are clorobiocin derivatives in which the alkyl side chains of the 4-hydroxy-benzoyl moieties have been modified (**Figure III.20**)[4]. They have been identified as very potent inhibitors of the growth of *S. aureus* ATCC 29213 and the methicillin-resistant *S. aureus* strain ATCC 43300 with MIC values of <0.06, equal to the parent compound clorobiocin[5]. Our study shows that these compounds are highly potent inhibitors of *S. aureus* DNA gyrase *in vitro*, with inhibitory concentrations of 1 nM. In contrast, the changes in the acyl side chains that provided increased DNA gyrase inhibitory activity, led to a lower inhibition of topoisomerase IV. It is also remarkable that the *E. coli* topoisomerase IV is more sensitive to the aminocoumarin antibiotics tested than the *S. aureus* enzyme. The results indicate that the presence of the 3``-MePC moiety and the natural occurring 3-dimethylallyl-4-hydroxy-benzoic acid side chain are of importance for topoisomerase IV inhibition.

compound name	IC$_{50}$ (µM)			
	E. coli		S. aureus	
	DNA gyrase	topo IV	DNA gyrase	topo IV
novobiocin	0.08	10	0.01	20
clorobiocin	0.03	3	0.006	10
coumermycin A$_1$	0.03	5	0.006	100
simocyclinone D8	0.1	>10	2	17
novclobiocin 101	0.3	3	0.05	35
novclobiocin 102	0.03	0.3	0.01	5
novclobiocin 103	0.1	>10	1	>50
novclobiocin 105	>100	>100	>100	>100
novclobiocin 217	0.006	8	0.001	>50
novclobiocin 225	0.006	8	0.001	>50

Table III.3:

Inhibitory activity of aminocoumarin antibiotics against DNA gyrase and topoisomerase IV from *Escherichia coli* and *Staphylococcus aureus*. The concentration [µM] of antibiotic that caused 50% inhibition (IC$_{50}$) of DNA gyrase supercoiling and topoisomerase IV decatenation is given. DNA gyrase and topoisomerase IV activities were determined in the presence of 700 and 100 mM potassium glutamate, respectively. Each assay was repeated at least twice and the IC$_{50}$ values were determined based on intensity of either the supercoiled DNA or decatenated DNA gel bands.

III.4. Inactivation of *cloHIJK* (encoding for Ring B) in the biosynthetic gene cluster of clorobiocin and heterologous expression of the modified cluster

Ring B of aminocoumarins has been linked to their toxicity. In order to replace the genuine Ring B of clorobiocin with less toxic analogous by mutasynthesis experiments, the biosynthesis of the genuine Ring B would need to be abolished.

Using RED/ET-mediated recombination, we replaced the coding sequence of *cloHIJK* with an apramycin resistance gene, resulting in cosmid clo-SA3 (**Figure III.22**). The resistance cassette was subsequently removed by restriction digestion and religation, using *Xba*I and *Spe*I sites introduced into the PCR primer sequence as described previously. This *cloHIJK* deficient gene cluster was integrated into the genome of S. coelicolor M512, resulting in S. coelicolor M512 (clo-SA4) (**Figure III.22**). Cultivation of this strain in clorobiocin production medium did not result in production of clorobiocin, while corresponding heterologous expression strains containing the intact gene cluster produced this antibiotic.

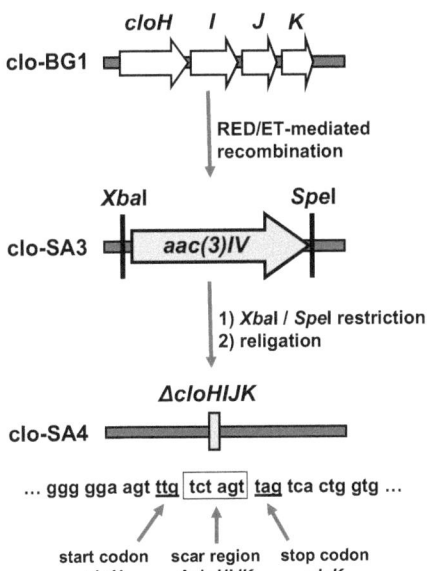

Figure III.22:
Gene deletion using the apramycin resistance cassette (*aac(3)IV*) from pIJ773 containing flanking *Xba*I and *Spe*I restriction sites.

However, feeding of Ring B (2 mg per 80 ml culture medium) restored production of clorobiocin, as confirmed by HPLC analysis and UV spectra comparison to an authentic reference compound (data not shown).

Therefore, the inactivation of *cloHIJK* had led to the abolishment of the production of ring B, but had not affected the subsequent steps of clorobiocin biosynthesis.

The only purchasable Ring B analogous 6-Amino-4-methyl-quinolin-2-ol (Matrix Scientific; **Figure III.23**) was fed under the same conditions as Ring B to *S. coelicolor* M512(clo-SA4). Until now, no product formation with this Ring B analogous could be observed by HPLC analysis.

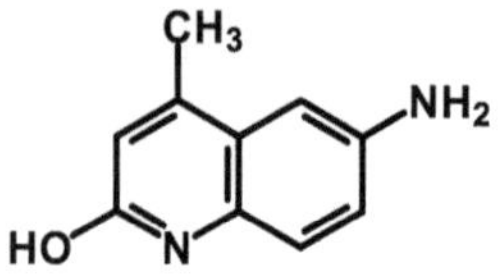

Figure III.23:
Ring B analogue 6-amino-4-methyl-quinolin-2-ol.

IV. DISCUSSION

IV.1. Generation and activity test of novclobiocin 401, a clorobiocin derivative containing the catechol moiety 3,4-dihydroxybenzoic acid

The present study provides an example for the rational design and production of a structurally modified antibiotic by a synthetic biology approach. The main aim was to exchange the structure of Ring A in clorobiocin by a catechol moiety, and in that way imitate a siderophore and facilitate the transport of the antibiotic inside the cell. The chemical synthesis of a clorobiocin derivative with a modified Ring A moiety is not straightforward, since a clorobiocin derivative lacking Ring A (e.g. containing only Rings B and C) can not be obtained from any producer strain, due to the fact that Ring A is the starter moiety for clorobiocin biosynthesis. Alternatively, the *de novo* chemical synthesis of an entire aminocoumarin antibiotic is a complicated multi-step procedure[63].

Our engineering strategy involved genes from four different organisms: i) the Gram-negative organism *E. coli* as the source of *ubiC*, which encodes an enzyme of an anabolic pathway; ii) the Gram-positive organism *Corynebactericum cyclohexanicum* as the source of *pobA*, encoding an enzyme of a catabolic pathway; iii) *Streptomyces roseochromogenes* as the source of the biosynthetic gene cluster of clorobiocin, which was modified by deletion of *cloQ*; and iv) *Streptomyces coelicolor* M512 as host for heterologous expression. *S. coelicolor* M512[38] is a derivate of the strain *S. coelicolor* A3(2). Strain M512 does not produce three of the genuine antibiotics of strain A3(2), i.e. methylenomycin, actinorhodin and undecylprodigiosine, which facilitates the detection and isolation of heterologously produced compounds. The modified clorobiocin gene cluster was stably integrated into the chromosome of this organism, and the synthetic operon containing *ubiC* and *pobA* under control of the strong constitutive *ermE** promoter was expressed from a replicative plasmid. This strategy resulted in the efficient production of the desired compound novclobiocin 401, a clorobiocin derivative in which the genuine Ring A is replaced with a 3,4-DHBA moiety.

The yield of novclobiocin 401 exceeded that of clorobiocin reported for both the wild type and for a heterologous producer strain[31]. Even without extensive optimization experiments, novclobiocin 401 represented a major compound in the culture extract, what facilitated the preparative isolation of this compound. The production of novclobiocin 401 did not result in any impairment of growth of the heterologous producer strain. Apparently, both the diversion of chorismate for the production of 3,4-DHBA and the accumulation of the potent gyrase inhibitor novclobiocin 401 were tolerated well. Self-resistance of the heterologous producer strain was expected, since the clorobiocin gene cluster, contained in clo-SA2, carries an aminocoumarin resistance gene[31].

In vitro investigation of the inhibitory effect of this compound on the DNA gyrase of *E. coli* and *S. aureus* showed no change of activity in comparison to the parent compound clorobiocin. This confirms that the structure of Ring A is of little importance for the interaction of aminocoumarin antibiotics with their principal target, and can be modified to introduce desirable motifs into the molecule. It should be noted that the IC_{50} of novclobiocin 401 against the gyrase of *E. coli* and *Staphylococcus aureus* is two to three orders of magnitude lower than that of modern fluoroquinolones[113], confirming the strong potency of aminocoumarins.

Novclobiocin 401, containing a catechol motif, showed higher antibacterial activity than its parent compound clorobiocin against *E. coli* mutants defective in the TolC-dependent efflux pump. When *tonB* was additionally deleted, the activity of novclobiocin 401 was reduced eightfold (as calculated from the MIC values), while the activity of clorobiocin remained unchanged. TonB supplies the energy for the catechol siderophore transporters Cir, Fiu and FepA, and therefore this result strongly suggests that these transporters are involved in the active import of novclobiocin 401.

Previous experiments showed that especially Cir and Fiu have a broad substrate specificity and can import catechol compounds of different structures[50]. Uptake of drug-siderophore conjugates by Cir and Fiu has been demonstrated previously, like for cephalosporin-catechol conjugates[50, 79]. However, in these examples the catechol moiety conjugated to the antibiotic added considerably to the molecular weight of the drug, and even the size of the catechol moiety often exceeded the size of the antibiotic moiety. In the present study, a pre-existing moiety (i.e. Ring A) of the antibiotic was replaced by a catechol moiety, and the molecular weight of the new catechol compound (645 Da) was lower than that of the parent compound clorobiocin (697 Da).

Though novclobiocin 401 was transported into *E. coli* cells under involvement of TonB-dependent transporters, it did not show higher activity than clorobiocin against wild-type and ΔentC mutants (with production of the natural siderophore enterobactin abolished) of

DISCUSSION

E. coli. The MIC values of clorobiocin and novclobiocin 401 against *E. coli* ΔentC mutants were 23 and 95 µg/ml, respectively. Under identical conditions, the MICs against ΔentC/ΔtolC double mutants were 12 and 6 µg/ml, respectively. These results suggest that novclobiocin 401 is more rapidly exported than clorobiocin by TolC-dependent drug efflux pumps. The introduction of a catechol moiety into the clorobiocin molecule therefore had the desired effect to facilitate active import by catechol siderophore transporters, but also the undesired effect to accelerate TolC-dependent efflux. Future attempts to improve the activity of aminocoumarins against Gram-negative organism may therefore aim at a reduced efflux, either by modification of the drug or by combination with an efflux pump inhibitor, and at a further improvement of active import, e.g. by inclusion of a 2,3-DHBA moiety, which is the catechol moiety present in enterobactin.

The generation of a clorobiocin derivative with a 2,3-DHBA moiety by combinatorial biosynthesis proved more complicated because of the requirement of its activation by adenylation prior to the acceptance by the aminocoumarin acyl ligase. We identified a suitable 2,3-DHBA AMP ligase from *B. subtilis*[75], but it failed to be functional in *S. coelicolor* M512. Further attempts to produce 2,3-DHBA in the same host by introducing a plasmid containing the necessary genes from the griseobactin gene cluster of *Streptomcyes griseus*[89] were also unsuccessful, probably because of inefficient expression of the genes from this construct.
We only obtained the desired 2,3-DHBA clorobiocin derivate when we fed 2,3-DHBA to the heterologous host *S. coelicolor* M1154 (improved heterologous expression strain derived from *S. coelicolor*[45]) containing clo-SA2 and the griseobactin-derived construct. We think that 2,3-DHBA, fed to the culture medium, was activated by an AMP-ligase endogenous in M1154, probably encoded by the first gene in the hypothetical coelibactin gene cluster[11], Sco7681. Although we have now obtained the desired compound, it is still to be tested *in vitro* and *in vivo* to see whether it is comparable imported as the 3,4-DHBA analogue novclobiocin 401.

The major focus of this work was an improved influx of aminocoumarins into Gram-negative bacteria. We have demonstrated that, at least in *E. coli*, it is the multidrug efflux-pump, dependent on the membrane channel TolC, which is responsible for the high level of resistance. The previously reported high sensitivity of novobiocin to *E. coli* mutants, affected in cardiolipin metabolism[116, 127], could well be due to an impaired function of the efflux pump in membranes diminished of that phospholipid. The increased activity shown by novclobiocin 401 against a TolC deficient *E. coli* strain is promising, and it must be noted that novclobiocin 401 (nor the 2,3-DHBA analogue) has not been tested *in vivo* against other bacteria yet.

All together, we are confident that the work presented in this thesis comprises an important advance in combinatorial biosynthesis of antibiotics and of aminocoumarins in particular towards an improvement of their bioactivity.

IV.2. Inhibition of DNA gyrase and topoisomerase IV of *Staphylococcus aureus* and *Escherichia coli* by aminocoumarin antibiotics

One part of the present work aimed at elucidating the contribution of different aminocoumarin structural elements to their inhibitory activity against DNA gyrase and topoisomerase IV of Gram-positive and Gram-negative bacteria, what would provide us with more knowledge for further rational improvement of this class of antibiotics.

We established an optimized protocol for supercoiling and decatenation assays, and we determined the inhibitory activity of four naturally occurring aminocoumarin antibiotics and of several novclobiocins, compounds derived from clorobiocin and novobiocin, against DNA gyrase and topoisomerase IV from *S. aureus* and *E. coli*.

In the supercoiling assay, *S. aureus* DNA gyrase requires the presence of high concentrations of K-Glu, whereas *E. coli* DNA gyrase had so far been assayed in the absence of this salt. For the first time, our study shows that the sensitivity of *E. coli* DNA gyrase towards the inhibition by aminocoumarins is considerably increased by K-Glu. Therefore, the concentration of K-Glu used in the assays needs to be considered when comparing the effect of inhibitors on *E. coli* and *S. aureus* DNA gyrase. The precise role of K-Glu in the catalytic mechanism of topoisomerases is unclear. Hiasa and co-workers[52], provided evidence that K-Glu is not required for the binding of DNA to the catalytic domain of DNA gyrase, but rather for its binding to the C-terminal domain of GyrA and the resulting wrapping of DNA which enables DNA gyrase to catalyse the supercoiling reaction.

Another theory is that DNA gyrase requires potassium for activity, and it has been shown that it stabilizes the GyrB subunit[108]. DNA Gyrase belongs to the GHKL family of enzymes (represented by gyrase, Hsp90, certain protein kinases, and the DNA mismatch protein

DISCUSSION

MutL), and crystallographic and biochemical studies have revealed a distinct binding site for monovalent cations like K^+ on these enzymes, which is important for catalytic activity[56]. This may explain the difference of the effects of K^+ and Na^+ on the type II topoisomerases. Assays with Na-Glu instead K-Glu showed no stimulatory effect for *S. aureus* DNA gyrase, and we even found an inhibitory effect over *E. coli* DNA gyrase. Since this enzyme does nor require K-Glu, the most feasible explanation is that the sodium added with Na-Glu outcompetes the potassium present in the reaction buffer as 24 mM KCl; this would explain the weak activity observed when adding only 100 mM Na-Glu, concentration at which a proportion of DNA gyrase could still be reached by potassium; and it would also explain the lack of activity of *E. coli* enzyme with increasing concentration of NaCl. The results obtained with competition assays between Na-Glu and KCl or K-Glu did not support this hypothesis and more experiments are required to give further insights into DNA gyrase requirements of potassium and glutamate.

Assays of inhibitory activity of aminocoumarin antibiotics reinforced the previous knowledge and provided further insights about the relevance of the different substituents on antibiotic activity. Out of the four naturally occurring aminocoumarin antibiotics, clorobiocin and coumermycin A_1 had the highest inhibitory activity against all the enzymes tested, and clorobiocin had the lowest IC_{50} toward topoisomerase IV of both *E. coli* and *S. aureus*. Manipulation of substituents[35] at position 3``-OH of noviose confirmed that the presence of MePC is essential for high inhibition of both DNA gyrase and topoisomerase IV. Furthermore, elimination of the methyl group at position 8`` of noviose rendered the compound (novclobiocin 105) completely inactive against all the enzymes tested, which indicate that hydrophobic contacts between this methyl group and a hydrophobic path of the enzymes play a prominent role[44, 64]. One of the most exciting results is the obtaining of clorobiocin and novobiocin derivates with stronger antibiotic activity than the natural compounds. In this way, modification of the Ring A, attached to the amino group of the coumarin ring, provided higher activity against DNA gyrase, particularly against *E. coli* enzyme, although with lost of activity against topoisomerase IV from both microorganisms. Therefore, our experiments suggest that the most active compounds against *S. aureus* and *E. coli* type II topoisomerases contain an 8``-$CH_3$ group, a MePC moiety at position 3`` and an 8`-CH_3 group.

These biochemical data suggest that *in vitro* DNA gyrase is the primary target of all investigated aminocoumarin antibiotics. Fujimoto-Nakamura and co-workers[42] have shown that cultivation of *S. aureus* in the presence of novobiocin results not only in the selection of mutants with altered DNA gyrase, but also in mutations of topoisomerase IV as a second step in resistance development. Therefore, topoisomerase IV may not be completely irrelevant as target of aminocoumarins in *S. aureus*, at least at higher antibiotic concentrations.

Another group of topoisomerases inhibitors are the synthetically generated fluoroquinolones, which interact with the GyrA and ParC subunits of DNA gyrase and topoisomerase IV respectively. Resistance against the fluoroquinolones is rapidly emerging, usually by mutations near the active site tyrosines of GyrA or ParC[29]. Agents that target type II topoisomerases in a different way than fluoroquinolones would offer a possibility to overcome this resistance while still exploiting the same validated target. The combination of fluoroquinolones with aminocoumarins, which are highly potent inhibitors of the GyrB and ParE subunits, may offer a strategy to provide effective antibacterial therapy with reduced risk of resistance development. *In vitro* studies have shown that *S. aureus* mutants that are simultaneously resistant to both fluoroquinolones and aminocoumarins arise only at very low frequency and might therefore not be selected if those agents were used as combination therapy[117]. These considerations may warrant the development and evaluation of new aminocoumarin antibiotics. The assay methods developed here can be useful in such approach.

V. REFERENCES

1 Althaus, I. W., Dolak, L., and Reusser, F. (1988) Coumarins as inhibitors of bacterial DNA gyrase, *J. Antibiotics* **41**: 373-376.

2 Alves, J. R., Pereira, A. C., Souza, M. C., Costa, S. B., Pinto, A. S., Mattos-Guaraldi, A. L. (2010) Iron-limited condition modulates biofilm formation and interaction with human epithelial cells of enteroaggregative *Escherichia coli* (EAEC), *J. Appl. Microbiol.* **108**: 246-255.

3 Anderle, C., Alt, S., Gulder, T., Bringmann, G., Kammerer, B., Gust, B., and Heide, L. (2007) Biosynthesis of clorobiocin: investigation of the transfer and methylation of the pyrrolyl-2-carboxyl moiety, *Arch. Microbiol.* **187**: 227-237.

4 Anderle, C., Hennig, S., Kammerer, B., Li, S. M., Wessjohann, L., Gust, B., and Heide, L. (2007) Improved mutasynthetic approaches for the production of modified aminocoumarin antibiotics, *Chem. Biol.* **14**: 955-967.

5 Anderle, C., Stieger, M., Burrell, M., Reinelt, S., Maxwell, A., Page, M., and Heide, L. (2008) Biological activities of novel gyrase inhibitors of the aminocoumarin class, *Antimicrob. Agents Chemother.* **52**: 1982-1990.

6 Arathoon, E. G., Hamilton, J. R., Hench, C. E., and Stevens, D. A. (1990) Efficacy of short courses of oral novobiocin-rifampin in eradicating carrier state of methicillin-resistant *Staphylococcus aureus* and in vitro killing studies of clinical isolates, *Antimicrob. Agents Chemother.* **34**: 1655-1659.

7 Augustus, A. M., Celaya, T., Husain, F., Humbard, M., and Misra, R. (2004) Antibiotic-sensitive TolC mutants and their suppressors, *J. Bacteriol.* **186**: 1851-1860.

8 Baba, T., Ara, T., Hasegawa, M., Takai, Y., Okumura, Y., Baba, M., et al. (2006) Construction of *Escherichia coli* K-12 in-frame, single-gene knockout mutants: the Keio collection, *Mol. Syst. Biol.* **2**: 2006-2008.

9 Baltz, R. H. (2008) Renaissance in antibacterial discovery from actinomycetes, *Curr. Opin. Pharmacol.* **8**: 557-563.

10 Bellon, S., Parsons, J. D., Wei, Y., Hayakawa, K., Swenson, L. L., Charifson, P. S., et al. (2004) Crystal structures of *Escherichia coli* topoisomerase IV ParE subunit (24 and 43 kilodaltons): a single residue dictates differences in novobiocin potency against topoisomerase IV and DNA gyrase, *Antimicrob. Agents Chemother.* **48**: 1856-1864.

11 Bentley, S. D., Chater, K. F., Cerdeno-Tarraga, A. M., Challis, G. L., Thomson, N. R., James, K. D. (2002) Complete genome sequence of the model actinomycete *Streptomyces coelicolor* A3(2), *Nature* **417**: 141-147.

REFERENCES

12 Benz, R., and Bauer, K. (1988) Permeation of hydrophilic molecules through the outer membrane of Gram-negative bacteria. Review on bacterial porins, *Europ. J. Biochem.* **176**: 1-19.

13 Berger, J., and Batcho, A. D. (1978) Coumarin-glycoside antibiotics. In: Antibiotics. Isolation, Separation and Purification. Weinstein, M. J., and Wagman, G. H. (eds). Amsterdam, Oxford, New York: Elsevier Scientific Publishing Company. 101-158.

14 Birch, A. J., Holloway, R. W., and Rickards, R. W. (1962) Biosynthesis of noviose, a branched-chain monosaccharide, *Biochim. Biophys. Acta* **57**: 148-145.

15 Blair, J. M., and Piddock, L. J. (2009) Structure, function and inhibition of RND efflux pumps in Gram-negative bacteria: an update, *Curr. Opin. in Microbiol.* **12**: 512-519.

16 Blanche, F., Cameron, B., Bernard, F. X., Maton, L., Manse, B., Ferrero, L. (1996) Differential behaviors of *Staphylococcus aureus* and *Escherichia coli* type II DNA topoisomerases, *Antimicrob. Agents Chemother.* **40**: 2714-2720.

17 Blattner, F. R., Plunkett, G., 3rd, Bloch, C. A., Perna, N. T., Burland, V., Riley, M., et al. (1997) The complete genome sequence of *Escherichia coli* K-12, *Science* **277**: 1453-1462.

18 Braun, V., and Hantke, K. (2001) Mechanisms of bacterial iron transport. *In* G. Winkelmann (ed.), Microbial transport systems, *Wiley, Weinheim-New York*: 289-311.

19 Bullock, W. O., Fernandez, J. M., and Short, J. M. (1987) XL1-Blue: a high efficiency plasmid transforming *recA Escherichia coliN* strain with beta-galactosidase selection, *Biotechniques* **5**: 376-379.

20 Bunton, C. A., Kenner, G. W., Robinson, M. J. T., and Webster, B. R. (1963) Experiments related to the biosynthesis of novobiocin and other coumarins, *Tetrahedron* **19**: 1001-1010.

21 Cane, D. E., Walsh, C. T., and Khosla, C. (1998) Harnessing the biosynthetic code: combinations, permutations, and mutations, *Science* **282**: 63-68.

22 Champoux, J. J. (2001) DNA topoisomerases: structure, function, and mechanism, *Annu. Rev. Biochem.* **70**: 369-413.

23 Chen, H., and Walsh, C. T. (2001) Coumarin formation in novobiocin biosynthesis: beta-hydroxylation of the aminoacyl enzyme tyrosyl-*S*-NovH by a cytochrome P450 NovI, *Chem. Biol.* **8**: 301-312.

24 Chu, B. C., Garcia-Herrero, A., Johanson, T. H., Krewulak, K. D., Lau, C. K., Peacock, R. S. (2010) Siderophore uptake in bacteria and the battle for iron with the host; a bird's eye view, *Biometals* **23**: 601-611.

REFERENCES

25 Cornelis, P. (2010) Iron uptake and metabolism in pseudomonads, *Appl. Microbiol. Biotechnol.* **86**: 1637-1645.

26 Crosa, J. H., and Walsh, C. T. (2002) Genetics and assembly line enzymology of siderophore biosynthesis in bacteria, *Microbiol. Mol. Biol. Rev.* **66**: 223-249.

27 Datsenko, K. A., and Wanner, B. L. (2000) One-step inactivation of chromosomal genes in *Escherichia coli* K-12 using PCR products, *Proc. Natl. Acad. Sci. USA* **97**: 6640-6645.

28 Doumith, M., Weingarten, P., Wehmeier, U. F., Salah-Bey, K., Benhamou, B., Capdevila, C. (2000) Analysis of genes involved in 6-deoxyhexose biosynthesis and transfer in *Saccharopolyspora erythraea*, *Mol. Gen. Genet.* **264**: 477-485.

29 Drlica, K., and Malik, M. (2003) Fluoroquinolones: action and resistance, *Curr. Top. Med. Chem.* **3**: 249-282.

30 Edwards, M. J., Flatman, R. H., Mitchenall, L. A., Stevenson, C. E., Le, T. B., Clarke, T. A. (2009) A crystal structure of the bifunctional antibiotic simocyclinone D8, bound to DNA gyrase, *Science* **326**: 1415-1418.

31 Eustáquio, A. S., Gust, B., Galm, U., Li, S. M., Chater, K. F., and Heide, L. (2005) Heterologous expression of novobiocin and clorobiocin biosynthetic gene clusters, *Appl. Environ. Microbiol.* **71**: 2452-2459.

32 Eustáquio, A. S., Gust, B., Luft, T., Li, S. M., Chater, K. F., and Heide, L. (2003) Clorobiocin biosynthesis in *Streptomyces*. Identification of the halogenase and generation of structural analogs, *Chem. Biol.* **10**: 279-288.

33 Eustáquio, A. S., Li, S. M., and Heide, L. (2005) NovG, a DNA-binding protein acting as a positive regulator of novobiocin biosynthesis, *Microbiology* **151**: 1949-1961.

34 Eustáquio, A. S., Luft, T., Wang, Z. X., Gust, B., Chater, K. F., Li, S. M., and Heide, L. (2003) Novobiocin biosynthesis: inactivation of the putative regulatory gene *novE* and heterologous expression of genes involved in aminocoumarin ring formation, *Arch. Microbiol.* **180**: 25-32.

35 Flatman, R. H., Eustáquio, A., Li, S. M., Heide, L., and Maxwell, A. (2006) Structure-activity relationships of aminocoumarin-type gyrase and topoisomerase IV inhibitors obtained by combinatorial biosynthesis, *Antimicrob. Agents Chemother.* **50**: 1136-1142.

36 Flatman, R. H., Howells, A. J., Heide, L., Fiedler, H. P., and Maxwell, A. (2005) Simocyclinone D8, an inhibitor of DNA gyrase with a novel mode of action, *Antimicrob. Agents Chemother.* **49**: 1093-1100.

37 Flinspach, K., Westrich, L., Kaysser, L., Siebenberg, S., Gomez-Escribano, J. P., Bibb, M., et al. (2010) Heterologous expression of the biosynthetic gene clusters of coumermycin A(1), clorobiocin and caprazamycins in genetically modified *Streptomyces coelicolor* strains, *Biopolymers* **93**: 823-832.

REFERENCES

38   Floriano, B., and Bibb, M. (1996) AfsR is a pleiotropic but conditionally required regulatory gene for antibiotic production in *Streptomyces coelicolor* A3(2), *Mol. Microbiol.* **21**: 385-396.

39   Freel Meyers, C. L., Oberthür, M., Anderson, J. W., Kahne, D., and Walsh, C. T. (2003) Initial characterization of novobiocic acid noviosyl transferase activity of NovM in biosynthesis of the antibiotic novobiocin, *Biochemistry* **42**: 4179-4189.

40   Freel Meyers, C. L., Oberthür, M., Xu, H., Heide, L., Kahne, D., and Walsh, C. T. (2004) Characterization of NovP and NovN: Completion of novobiocin biosynthesis by sequential tailoring of the noviosyl ring, *Angew. Chem. Int. Ed. Engl.* **43**: 67-70.

41   Fujii, T., and Kaneda, T. (1985) Purification and properties of NADH/NADPH-dependent p-hydroxybenzoate hydroxylase from *Corynebacterium cyclohexanicum*, *Europ. J. Biochem.* **147**: 97-104.

42   Fujimoto-Nakamura, M., Ito, H., Oyamada, Y., Nishino, T., and Yamagishi, J. (2005) Accumulation of mutations in both *gyrB* and *parE* genes is associated with high-level resistance to novobiocin in *Staphylococcus aureus*, *Antimicrob. Agents Chemother.* **49**: 3810-3815.

43   Galm, U., Dessoy, M. A., Schmidt, J., Wessjohann, L. A., and Heide, L. (2004) *In vitro* and *in vivo* production of new aminocoumarins by a combined biochemical, genetic, and synthetic approach, *Chem. Biol.* **11**: 173-183.

44   Galm, U., Heller, S., Shapiro, S., Page, M., Li, S. M., and Heide, L. (2004) Antimicrobial and DNA gyrase-inhibitory activities of novel clorobiocin derivatives produced by mutasynthesis, *Antimicrob. Agents Chemother.* **48**: 1307-1312.

45   Gomez-Escribano, J. P., and Bibb, M. J. (2010) Engineering *Streptomyces coelicolor* for heterologous expression of secondary metabolite gene clusters, *Microb. Biotechnol.* **4**: 207-215.

46   Gust, B. (2009) Cloning and analysis of natural product pathways, *Meth. Enzymol.* **458**: 159-180.

47   Gust, B., Challis, G. L., Fowler, K., Kieser, T., and Chater, K. F. (2003) PCR-targeted *Streptomyces* gene replacement identifies a protein domain needed for biosynthesis of the sesquiterpene soil odor geosmin, *Proc. Natl. Acad. Sci. USA* **100**: 1541-1546.

48   Heide, L. (2009) Genetic engineering of antibiotic biosynthesis for the generation of new aminocoumarins, *Biotechnol. Adv.* **27**: 1006-1014.

49   Heide, L. (2009) The aminocoumarins: biosynthesis and biology, *Nat. Product Reports* **26**: 1241–1250.

REFERENCES

50 Heinisch, L., Wittmann, S., Stoiber, T., Scherlitz-Hofmann, I., Ankel-Fuchs, D., and Mollmann, U. (2003) Synthesis and biological activity of tris- and tetrakiscatecholate siderophores based on poly-aza alkanoic acids or alkylbenzoic acids and their conjugates with beta-lactam antibiotics, *Arzneimittel-Forschung* **53**: 188-195.

51 Hesketh, A., Kock, H., Mootien, S., and Bibb, M. (2009) The role of *absC*, a novel regulatory gene for secondary metabolism, in zinc-dependent antibiotic production in *Streptomyces coelicolor* A3(2), *Mol. Microbiol.* **74**: 1427-1444.

52 Hiasa, H., Shea, M. E., Richardson, C. M., and Gwynn, M. N. (2003) *Staphylococcus aureus* gyrase-quinolone-DNA ternary complexes fail to arrest replication fork progression *in vitro*. Effects of salt on the DNA binding mode and the catalytic activity of *S. aureus* gyrase, *J. Biol. Chem.* **278**: 8861-8868.

53 Hopwood, D. A. (2009) Complex enzymes in microbial natural product biosynthesis; Part A: overview articles and peptides., *Meth. Enzymol.* **458**.

54 Hopwood, D. A. (2009) Complex enzymes in microbial natural product biosynthsis. part B: polyketides, aminocoumarins and carbohydrates., *Meth. Enzymol.* **495**.

55 Hopwood, D. A., Malpartida, F., Kieser, H. M., Ikeda, H., Duncan, J., Fujii, I. (1985) Production of 'hybrid' antibiotics by genetic engineering, *Nature* **314**: 642-644.

56 Hu, X., Machius, M., and Yang, W. (2003) Monovalent cation dependence and preference of GHKL ATPases and kinases, *FEBS Letters* **544**: 268-273.

57 Huang, Y., Zhao, K. X., Shen, X. H., Jiang, C. Y., and Liu, S. J. (2008) Genetic and biochemical characterization of a 4-hydroxybenzoate hydroxylase from *Corynebacterium glutamicum*, *Appl. Microbiol. Biotechnol.* **78**: 75-83.

58 Kammerer, B., Kahlich, R., Laufer, S., Li, S. M., Heide, L., and Gleiter, C. H. (2004) Mass spectrometric pathway monitoring of secondary metabolites: systematic analysis of culture extracts of *Streptomyces* species, *Anal. Biochem.* **335**: 17-29.

59 Kieser, T., Bibb, M. J., Buttner, M. J., Chater, K. F., and Hopwood, D. A. (2000) Practical Streptomyces Genetics, 2nd edition. John Innes Foundation, Norwich, UK.

60 Kominek, L. A. (1972) Biosynthesis of novobiocin by *Streptomyces niveus*, *Antimicrob. Agents Chemother.* **1**: 123-134.

61 Kumarasamy, K. K., Toleman, M. A., Walsh, T. R., Bagaria, J., Butt, F., Balakrishnan, R. (2010) Emergence of a new antibiotic resistance mechanism in India, Pakistan, and the UK: a molecular, biological, and epidemiological study, *Lancet. Infect. Dis.* **10**: 597-602.

62 Laemmli, U. K. (1970) Cleavage of structural proteins during the assembly of the head of bacteriophage T4, *Nature* **227**: 680-685.

REFERENCES

63   Laurin, P., Ferroud, D., Klich, M., Dupuis-Hamelin, C., Mauvais, P., Lassaigne, P., et al. (1999) Synthesis and *in vitro* evaluation of novel highly potent coumarin inhibitors of gyrase B, *Bioorg. Med. Chem. Lett.* **9**: 2079-2084.

64   Lewis, R. J., Singh, O. M. P., Smith, C. V., Skarzynski, T., Maxwell, A., Wonacott, A. J., and Wigley, D. B. (1996) The nature of inhibition of DNA gyrase by the coumarins and the cyclothialidines revealed by X-ray crystallography, *EMBO J.* **15**: 1412-1420.

65   Li, S. M., and Heide, L. (2005) New aminocoumarin antibiotics from genetically engineered *Streptomyces* strains, *Curr. Med. Chem.* **12**: 419-427.

66   Li, S. M., and Heide, L. (2006) The biosynthetic gene clusters of aminocoumarin antibiotics, *Planta Medica* **72**: 1093-1099.

67   Li, S. M., Westrich, L., Schmidt, J., Kuhnt, C., and Heide, L. (2002) Methyltransferase genes in *Streptomyces rishiriensis*: new coumermycin derivatives from gene-inactivation experiments, *Microbiology* **148**: 3317-3326.

68   Lorico, A., Rappa, G., and Sartorelli, A. C. (1992) Novobiocin-induced accumulation of etoposide (VP-16) in WEHI-3B D+ leukemia cells, *Int. J. Cancer* **52**: 903-909.

69   Luft, T., Li, S. M., Scheible, H., Kammerer, B., and Heide, L. (2005) Overexpression, purification and characterization of SimL, an amide synthetase involved in simocyclinone biosynthesis, *Arch. Microbiol.* **183**: 277-285.

70   Luria, S. E., and Burrous, J. W. (1957) Hybridization between *Escherichia coli* and *Shigella*, *J. Bacteriology* **74**: 461-476.

71   MacNeil, D. J., Gewain, K. M., Ruby, C. L., Dezeny, G., Gibbons, P. H., and MacNeil, T. (1992) Analysis of *Streptomyces avermitilis* genes required for avermectin biosynthesis utilizing a novel integration vector, *Gene* **111**: 61-68.

72   Mancy, D., Ninet, L., and Preudi Homme, J. (1974) U.S. Patent 3793147.

73   Marcu, M. G., Schulte, T. W., and Neckers, L. (2000) Novobiocin and related coumarins and depletion of heat shock protein 90-dependent signaling proteins, *J. Nat. Cancer Institute* **92**: 242-248.

74   Maxwell, A., and Lawson, D. M. (2003) The ATP-binding site of type II topoisomerases as a target for antibacterial drugs, *Curr. Top. Med. Chem.* **3**: 283-303.

75   May, J., Wendrich, T., and Marahiel, M. (2001) The *dhb* Operon of *Bacillus subtilis* encodes the biosynthetic template for the catecholic siderophore 2,3-dihydroxybenzoate-glycine-threonine trimeric ester bacillibactin, *Biolog. Chem.* **276 (10)**: 7209-7217.

76 McDaniel, R., Thamchaipenet, A., Gustafsson, C., Fu, H., Betlach, M., and Ashley, G. (1999) Multiple genetic modifications of the erythromycin polyketide synthase to produce a library of novel "unnatural" natural products, *Proc. Natl. Acad. Sci. USA* **96**: 1846-1851.

77 Meganathan, R. (2001) Ubiquinone biosynthesis in microorganisms, *FEMS Microbiol. Lett.* **203**: 131-139.

78 Merkens, H., Beckers, G., Wirtz, A., and Burkovski, A. (2005) Vanillate metabolism in *Corynebacterium glutamicum*, *Curr. Microbiol.* **51**: 59-65.

79 Möllmann, U., Heinisch, L., Bauernfeind, A., Kohler, T., and Ankel-Fuchs, D. (2009) Siderophores as drug delivery agents: application of the "Trojan Horse" strategy, *Biometals* **22**: 615-624.

80 Morgan-Linnell, S. K., Hiasa, H., Zechiedrich, L., and Nitiss, J. L. (2007) Assessing sensitivity to antibacterial topoisomerase II inhibitors, *Curr. Protoc. Pharmacol.* **39**: 3.13.11-13.13.26.

81 Nagakubo, S., Nishino, K., Hirata, T., and Yamaguchi, A. (2002) The putative response regulator BaeR stimulates multidrug resistance of *Escherichia coli* via a novel multidrug exporter system, MdtABC, *J. Bacteriol.* **184**: 4161-4167.

82 Neilands, J. B. (1995) Siderophores: structure and function of microbial iron transport compounds, *J. Biol. Chem.* **270**: 26723-26726.

83 Nikaido, H. (1976) Outer membrane of *Salmonella typhimurium*. Transmembrane diffusion of some hydrophobic substances, *Biochim. Biophys. Acta* **433**: 118-132.

84 Nikaido, H. (1998) Antibiotic resistance caused by Gram-negative multidrug efflux pumps, *Clinic. Infect. Diseases* **27** Suppl. 1: 32-41.

85 Nikaido, H., and Vaara, M. (1985) Molecular basis of bacterial outer membrane permeability, *Microbiol. Reviews* **49**: 1-32.

86 Oblak, M., Kotnik, M., and Solmajer, T. (2007) Discovery and development of ATPase inhibitors of DNA gyrase as antibacterial agents, *Curr. Med. Chem.* **14**: 2033-2047.

87 Oppegard, L. M., Hamann, B. L., Streck, K. R., Ellis, K. C., Fiedler, H. P., Khodursky, A. B., and Hiasa, H. (2009) In vivo and in vitro patterns of the activity of simocyclinone D8, an angucyclinone antibiotic from *Streptomyces antibioticus*, *Antimicrob. Agents Chemother.* **53**: 2110-2119.

88 Pan, X. S., and Fisher, L. M. (1999) *Streptococcus pneumoniae* DNA gyrase and topoisomerase IV: overexpression, purification, and differential inhibition by fluoroquinolones, *Antimicrob. Agents Chemother.* **43**: 1129-1136.

89 Patzer, S. I., and Braun, V. (2010) Gene cluster involved in the biosynthesis of griseobactin, a catechol-peptide siderophore of *Streptomyces sp.* ATCC 700974, *J. Bacteriol.* **192**: 426-435.

REFERENCES

90 Peng, H., and Marians, K. J. (1993) Decatenation activity of topoisomerase IV during *oriC* and pBR322 DNA replication *in vitro*, *Proc. Natl. Acad. Sci. USA* **90**: 8571-8575.

91 Pojer, F., Kahlich, R., Kammerer, B., Li, S. M., and Heide, L. (2003) CloR, a bifunctional non-heme iron oxygenase involved in clorobiocin biosynthesis, *J. Biol. Chem.* **278**: 30661-30668.

92 Pojer, F., Li, S. M., and Heide, L. (2002) Molecular cloning and sequence analysis of the clorobiocin biosynthetic gene cluster: new insights into the biosynthesis of aminocoumarin antibiotics, *Microbiology* **148**: 3901-3911.

93 Pojer, F., Wemakor, E., Kammerer, B., Chen, H., Walsh, C. T., Li, S. M., and Heide, L. (2003) CloQ, a prenyltransferase involved in clorobiocin biosynthesis, *Proc. Natl. Acad. Sci. USA* **100**: 2316-2321.

94 Ponte-Sucre, A. (2009) ABC transporters in microorganisms – research, innovation and value as targets against drug resistance. Caister Academic Press, UK.

95 Rappa, G., Lorico, A., and Sartorelli, A. C. (1992) Potentiation by novobiocin of the cytotoxic activity of etoposide (VP-16) and teniposide (VM-26), *Int. J. Cancer* **51**: 780-787.

96 Rappa, G., Murren, J. R., Johnson, L. M., Lorico, A., and Sartorelli, A. C. (2000) Novobiocin-induced VP-16 accumulation and MRP expression in human leukemia and ovarian carcinoma cells, *Anticancer Drug Des.* **15**: 127-134.

97 Rappa, G., Shyam, K., Lorico, A., Fodstad, O., and Sartorelli, A. C. (2000) Structure-activity studies of novobiocin analogs as modulators of the cytotoxicity of etoposide (VP-16), *Oncol. Res.* **12**: 113-119.

98 Reusser, F., and Dolak, L. A. (1986) Novenamine is the active moiety in novobiocin, *J. Antibiotics* **39**: 272-274.

99 Saiki, A. Y. C., Shen, L. L., Chen, C. M., Baranowski, J., and Lerner, C. G. (1999) DNA cleavage activities of *Staphylococcus aureus* gyrase and topoisomerase IV stimulated by quinolones and 2-pyridones, *Antimicrob. Agents Chemother.* **43**: 1574-1577.

100 Sambrook, J., and Russell, D. W. (2001) Molecular Cloning. A Laboratory Manual. New York Cold Spring Harbor Laboratory Press.

101 Schimana, J., Fiedler, H. P., Groth, I., Süssmuth, R., Beil, W., Walker, M., and Zeeck, A. (2000) Simocyclinones, novel cytostatic angucyclinone antibiotics produced by *Streptomyces antibioticus* Tü 6040. I. Taxonomy, fermentation, isolation and biological activities, *J.Antibiotics* **53**: 779-787.

102 Schmutz, E., Mühlenweg, A., Li, S. M., and Heide, L. (2003) Resistance genes of aminocoumarin producers: Two type II topoisomerase genes confer resistance against coumermycin A_1 and clorobiocin, *Antimicrob. Agents Chemother.* **47**: 869-877.

REFERENCES

103 Schmutz, E., Steffensky, M., Schmidt, J., Porzel, A., Li, S. M., and Heide, L. (2003) An unusual amide synthetase (CouL) from the coumermycin A_1 biosynthetic gene cluster from *Streptomyces rishiriensis* DSM 40489, *Europ. J. Biochem.* **270**: 4413-4419.

104 Schoeffler, A. J., and Berger, J. M. (2005) Recent advances in understanding structure-function relationships in the type II topoisomerase mechanism, *Biochem. Soc. Trans.* **33**: 1465-1470.

105 Shima, J., Penyige, A., and Ochi, K. (1996) Changes in patterns of ADP-ribosylated proteins during differentiation of *Streptomyces coelicolor* A3(2) and its development mutants, *J. Bacteriol.* **178**: 3785-3790.

106 Siebert, M., Bechthold, A., Melzer, M., May, U., Berger, U., Schröder, G. (1992) Ubiquinone biosynthesis. Cloning of the genes coding for chorismate pyruvate-lyase and 4-hydroxybenzoate octaprenyl transferase from *Escherichia coli*, *FEBS Lett.* **307**: 347-350.

107 Silver, L. L. (2007) Multi-targeting by monotherapeutic antibacterials, *Nat. Rev. Drug Discov.* **6**: 41-55.

108 Sissi, C., Marangon, E., Chemello, A., Noble, C. G., Maxwell, A., and Palumbo, M. (2005) The effects of metal ions on the structure and stability of the DNA gyrase B protein, *J. Mol. Biol.* **353**: 1152-1160.

109 Sohng, J. K., Oh, T. J., Lee, J. J., and Kim, C. G. (1997) Identification of a gene cluster of biosynthetic genes of rubradirin substructures in *S. achromogenes var. rubradiris* NRRL3061, *Mol. Cells* **7**: 674-681.

110 Steffensky, M., Li, S. M., and Heide, L. (2000) Cloning, overexpression, and purification of novobiocic acid synthetase from *Streptomyces spheroides* NCIMB 11891, *J. Biol. Chem.* **275**: 21754-21760.

111 Steffensky, M., Mühlenweg, A., Wang, Z. X., Li, S. M., and Heide, L. (2000) Identification of the novobiocin biosynthetic gene cluster of *Streptomyces spheroides* NCIB 11891, *Antimicrob. Agents Chemother.* **44**: 1214-1222.

112 Strahilevitz, J., Robicsek, A., and Hooper, D. C. (2006) Role of the extended alpha4 domain of *Staphylococcus aureus* gyrase A protein in determining low sensitivity to quinolones, *Antimicrob. Agents Chemother.* **50**: 600-606.

113 Takei, M., Fukuda, H., Kishii, R., and Hosaka, M. (2001) Target preference of 15 quinolones against *Staphylococcus aureus*, based on antibacterial activities and target inhibition, *Antimicrob. Agents Chemother.* **45**: 3544-3547.

114 Tamaki, S., Sato, T., and Matsuhashi, M. (1971) Role of lipopolysaccharides in antibiotic resistance and bacteriophage adsorption of *Escherichia coli* K-12, *J. Bacteriol.* **105**: 968-975.

REFERENCES

115 Tanaka, M., Onodera, Y., Uchida, Y., Sato, K., and Hayakawa, I. (1997) Inhibitory activities of quinolones against DNA gyrase and topoisomerase IV purified from *Staphylococcus aureus*, *Antimicrob. Agents Chemother.* **41**: 2362-2366.

116 Tropp, B. E., Ragolia, L., Xia, W., Dowhan, W., Milkman, R., Rudd, K. E. (1995) Identity of the *Escherichia coli cls* and *nov* genes, *J Bacteriol.* **177**: 5155-5157.

117 Vickers, A. A., O'Neill, A. J., and Chopra, I. (2007) Emergence and maintenance of resistance to fluoroquinolones and coumarins in *Staphylococcus aureus*: predictions from *in vitro* studies, *J. Antimicrob. Chemother.* **60**: 269-273.

118 Viitanen, P. V., Devine, A. L., Khan, M. S., Deuel, D. L., Van Dyk, D. E., and Daniell, H. (2004) Metabolic engineering of the chloroplast genome using the *Echerichia coli ubiC* gene reveals that chorismate is a readily abundant plant precursor for p-hydroxybenzoic acid biosynthesis, *Plant Physiol.* **136**: 4048-4060.

119 Walsh, C. T. (2002) Combinatorial biosynthesis of antibiotics: challenges and opportunities, *Chembiochem* **3**: 124-134.

120 Walsh, C. T. (2003) Where will new antibiotics come from?, *Nat. Rev. Microbiol.* **1**: 65-70.

121 Wang, Z. X., Li, S. M., and Heide, L. (2000) Identification of the coumermycin A_1 biosynthetic gene cluster of *Streptomyces rishiriensis* DSM 40489, *Antimicrob. Agents Chemother.* **44**: 3040-3048.

122 Watve, M. G., Tickoo, R., Jog, M. M., and Bhole, B. D. (2001) How many antibiotics are produced by the genus *Streptomyces* ?, *Arch. Microbiol.* **176**: 386-390.

123 Wiegand, I., Hilpert, K., and Hancock, R. E. (2008) Agar and broth dilution methods to determine the minimal inhibitory concentration (MIC) of antimicrobial substances, *Nat. Protoc.* **3**: 163-175.

124 Wohlleben, W., and Pelzer, S. (2002) New compounds by combining "modern" genomics and "old-fashioned" mutasynthesis, *Chem. Biol.* **9**: 1163-1164.

125 Wright, F., and Bibb, M. J. (1992) Codon usage in the G+C-rich *Streptomyces* genome, *Gene* **113**: 55-65.

126 Xu, H., Kahlich, R., Kammerer, B., Heide, L., and Li, S. M. (2003) CloN2, a novel acyltransferase involved in the attachment of the pyrrole-2-carboxyl moiety to the deoxysugar of clorobiocin, *Microbiology* **149**: 2183-2191.

127 Yethon, J. A., and Whitfield, C. (2001) Purification and characterization of WaaP from *Escherichia coli*, a lipopolysaccharide kinase essential for outer membrane stability, *J. Biol. Chem.* **276(8)**: 5498-5504.

128 Zhang, W., Heemstra, J., Walsh, C., and Imker, H. (2010) Activation of the pacidamycin PacL adenylation domain by MbtH-like proteins, *Biochemistry* **49(46)**: 9946-9947.

I want morebooks!

Buy your books fast and straightforward online - at one of world's fastest growing online book stores! Environmentally sound due to Print-on-Demand technologies.

Buy your books online at
www.morebooks.shop

Kaufen Sie Ihre Bücher schnell und unkompliziert online – auf einer der am schnellsten wachsenden Buchhandelsplattformen weltweit! Dank Print-On-Demand umwelt- und ressourcenschonend produziert.

Bücher schneller online kaufen
www.morebooks.shop

KS OmniScriptum Publishing
Brivibas gatve 197
LV-1039 Riga, Latvia
Telefax: +371 686 204 55

info@omniscriptum.com
www.omniscriptum.com

Printed by Books on Demand GmbH, Norderstedt / Germany